CliffsNotes® Statistics and Probability Common Core Quick Review

CliffsNotes® Statistics and Probability Common Core Quick Review

By Malihe Alikhani, M.S.

About the Author
Malihe Alikhani, M.S., is an author and educator who earned her advanced degree in mathematics from the University of Hawaii at Manoa. Currently a computer science-machine learning Ph.D. candidate at Rutgers University, she has taught statistics and mathematics at San Diego Community College and San Diego State University.

Editorial
Executive Editor: Greg Tubach
Senior Editor: Christina Stambaugh
Production Editor: Erika West
Copy Editor: Lynn Northrup
Technical Editors: Mary Jane Sterling and Tom Page
Proofreader: Pamela Weber-Leaf
Indexer: Potomac Indexing, LLC

CliffsNotes® Statistics and Probability Common Core Quick Review

Library of Congress Control Number: 2016957653
ISBN: 978-1-957671-18-5

Printed in the United States of America
DOC 10 9 8 7 6 5 4 3 2 1

Table of Contents

INTRODUCTION

Statistics is a powerful mathematical tool for designing an experiment, collecting data, describing data, and interpreting data. We are accustomed to seeing statistics every day—weather reports, political polls, advertising claims, classroom grade reports, and so on. However, many people find statistics confusing and intimidating. *CliffsNotes Statistics and Probability Common Core Quick Review* is intended to clarify some of the fundamentals of statistical reasoning to make informed decisions about raw data in everyday scenarios. The format is simple, and there are many examples to help you translate abstract data into concrete applications.

Your knowledge of arithmetic operations is fundamental to the topics presented in this book. If you feel that your knowledge of arithmetic requires further review, refer to *CliffsNotes Basic Math and Pre-Algebra Quick Review.*

Connecting Statistics to the Common Core State Standards for Mathematics (CCSSM)

CliffsNotes Statistics and Probability Common Core Quick Review topics are aligned with the Common Core State Standards for Mathematics (CCSSM)—the skills you should know and be able to perform. Statistics and probability concepts presented in this book make connections to other Common Core mathematics topics. Our goal is to help you become an advanced critical thinker so that you can make logical connections between the topics presented in the book and topics related to the CCSSM. This guide gives you procedural approaches to analyze numerical and categorical data in your daily life and to make meaningful interpretations that will support your conclusions in real-world scenarios.

Use the following guidelines in the practice of Common Core Mathematics statistics and probability topics:

- Make sense of problems, persevere in solving problems, and continue to monitor and evaluate your progress.
- Reason abstractly and quantitatively to describe relationships between variables of categorical and quantitative data.
- Construct viable arguments and analyze your reasoning as you deconstruct, organize, and make sense of problems.
- Justify your plausible conclusions and clarify your inferences from the categorical and quantitative data.
- Model with mathematics by using probability to evaluate outcomes of decisions.
- Use appropriate electronic tools (e.g., electronic calculators, software, or any other technology) to compute and interpret statistical data.
- Attend to precision as you carefully check your own reasoning.
- Look for and make use of structural patterns in the data presented in graphs, charts, and tables from sample surveys, experiments, and studies.
- Look for and express statistical reasoning regularly to improve your proficiency.

Why You Need This Book

Can you answer yes to any of these questions?

- Do you need to review the fundamentals of statistics quickly?
- Do you need a course supplement to statistics?
- Do you need a concise, comprehensive reference for statistics?
- Do you need to prepare for a statistics exam?

If so, then *CliffsNotes Statistics and Probability Common Core Quick Review* is for you!

How to Use This Book

You can use this book in any way that fits your personal style for study and review—you decide what works best with your needs. You can read the book from cover to cover or just look up the information you want and put it back on the shelf for later. Most people find it useful to follow the recommended sequence of topics in chapters 1–8.

Here are just a few ways you can search for topics:

- Look for areas of interest in the book's table of contents or use the index to find specific topics.
- Flip through the book looking for subject areas by heading.
- Get a glimpse of what you'll gain from a chapter by reading through the "Chapter Check-In" at the beginning of each chapter.
- Use the "Chapter Check-Out" at the end of each chapter to gauge your grasp of the important information you need to know.
- Test your knowledge more completely in the "Review Questions" (pp. 133–142).
- Look in the glossary (pp. 143–149) to find key terms fast. This book defines new terms and concepts where they first appear in the chapter. If a word is boldfaced, you can find a more complete definition in the book's glossary.
- Flip through the book until you find what you're looking for—we organized this book to gradually build on key concepts.

Hundreds of Practice Questions Online!

Go to CliffsNotes.com for hundreds of additional Statistics and Probability Common Core practice questions to help you prepare for your next quiz or test. The questions are organized by this book's chapter sections, so it is easy to use the book and then quiz yourself online to make sure you know the subject. Visit CliffsNotes.com to test yourself anytime and find other free homework help.

Chapter 1
OVERVIEW

Chapter Check-In

- Understanding the different types of statistics
- Learning about statistical variables
- Making an inference
- Evaluating several uses for statistical reasoning—making predictions, comparing results, and probability
- Avoiding common statistics mistakes

Common Core Standard: Interpreting Categorical and Quantitative Data

Summarize, represent, and interpret data on a single count or measurement variable. Summarize, represent, and interpret data on two categorical and quantitative variables (S.ID). Interpret linear models.

tatistics is the name given to the mathematical science of organizing and analyzing numbers—numerical facts, figures, and other information. Statistics is about understanding data. Data consist of information about objects that can be measured, called statistical variables. Common Core statistics draws upon your ability to organize, analyze, and interpret numerical data so that reliable patterns can provide a clear picture about events and people in the world.

Types of Statistics

Statistics is divided into two areas: descriptive statistics and inferential

Descriptive statistics

Reports of baseball batting averages, government surpluses or deficits, weather pattern trends, and so forth are often called statistical reports. To be precise, these numbers are **descriptive statistics** because they are numerical data that *describe* phenomena in our world. Descriptive statistics are as simple as the number of children in each family along a city block, or as complex as the annual report released from the U.S. Department of the Treasury about foreign investments.

nferential statistics

Statistics is also a *method,* a way of working with numbers to answer puzzling questions and draw conclusions about both human and nonhuman phenomena. **Inferential statistics** are not just concerned with describing data; they are used to make *inferences* about the data collected. Questions that can be answered using the inferential method of statistics are many and varied, including these: Which of several techniques is best for teaching reading to third graders? Will a new medicine be more effective than the old one? Can you expect it to rain tomorrow? What is the probable outcome of the next presidential election? Which assembly-line process produces fewer faulty carburetors? How can a polling organization make an accurate prediction of a national election by questioning only a few thousand voters?

For our purposes, statistics is both a collection of numbers and/or pictures a process: the art and science of making accurate guesses about outcomes involving numbers. In this chapter, we are concerned with two ways of representing descriptive statistics: numerical and pictorial.

Numerical statistics

Numerical statistics are numbers, but clearly, some numbers are more meaningful than others. For example, if you are offered a purchase price of $1 for an automobile on the condition that you also buy a second automobile, the price of the second automobile would be a major consideration (its price could be $1,000,000 or $1,000); thus, the average—or —of the two prices would be the important statistic. (The different meaningful numerical statistics are discussed in Chapter 3.)

Pictorial statistics

types of graphs are used for quantitative and categorical variables. (The more commonly used graphic displays in statistics are discussed in Chapter 2.)

So, fundamentally, the goals of statistics are:

To describe variables and data

To make accurate inferences about groups based upon incomplete information

Statistical Variables

(plural) consist of information about the measurement of statistical variables. A **variable** is the distinctive characteristic of an object that can be measured. A variable can differ from one object to another, such as gender, weight, size, time of day, score, number of objects, and so on. There are two types of variables: quantitative and categorical. **Quantitative variables** are variables that can be numerically measured or described such as the height of a pet. **Categorical variables** have values that are categories, such as the type of a pet. The data for these variables are usually counts or frequencies of the numbers for each category.

Another important way to look at statistical variables is to describe them as dependent or independent variables. A **dependent variable** is affected by the changes in other variables. An **independent variable,** on the other hand, is a variable that is independent of changes of other variables. This classification is particularly helpful for solving **probability** problems. We will explain this in detail in Chapter 4.

Making an Inference

Making accurate guesses requires groundwork. The statistician must do the following in order to make an educated, better-than-chance hunch:

Gather data (numerical information).

Organize the data (sometimes graphically).

Analyze the data (using tests of significance and so forth).

This book shows you how to follow these procedures and to use your analysis to draw an **inference**—an educated statistical guess—to solve a

Making Predictions

Suppose that you decide to sell commemorative T-shirts at your town's centennial picnic. You know that you can make a tidy profit, but only if you can sell most of your supply of shirts because your supplier will not buy any of them back. How many shirts can you reasonably plan on selling?

Your first question, of course, is this: How many people will be attending the picnic? Suppose you know that 100,000 tickets to the event have been sold. How many T-shirts should you purchase from your supplier to sell on the day of the event? 10,000? 50,000? 100,000? 150,000? How many of each size—small, medium, large, extra-large? And the important question: How many T-shirts must you sell in order to make some profit for your time *not* be left with an inventory of thousands of unsold shirts?

you buy your inventory of T-shirts, you need to have an accurate idea of just how many ticket holders will want to purchase centennial T-shirts and which sizes they will want. But, obviously, you have neither the time nor the resources to ask all 100,000 people whether they plan to purchase commemorative T-shirts. If, however, you could locate a small number of those ticket holders—for example, 100—and get an accurate count of how many of those 100 would purchase a T-shirt, you would have a better idea of how many of the 100,000 ticket holders would be willing to buy one.

That is, of course, if the 100 ticket holders that you asked (called the) are not too different in their intentions to purchase T-shirts from the total 100,000 ticket holders (called the **population**). If the sample is indeed representative (typical) of the population, you could expect about the same percentage of T-shirt sales (and sizes) for the population as for the sample, all things being equal. So, if 50 of your sample of 100 people say they cannot wait to plunk down $10 for a centennial T-shirt, it would be reasonable to expect that you would sell about 50,000 T-shirts to your population of 100,000. (At just $1 profit per T-shirt, that is $50,000!)

But before you start shopping to buy a yacht with your profits, remember that this prediction of total T-shirt sales relies heavily upon the sample being representative (similar to the population), which may not necessarily be the case with your sample. You may have inadvertently selected a sample that has more expendable income or a greater proportion of souvenir T-shirt enthusiasts or who knows what else. Are you reasonably

You can see why choosing a random sample is a critical part of the process of statistics and why this book includes a chapter on random sampling (Chapter 5). Even with careful sampling methods, our conclusions are still educated guesses. This is because one sample does not perfectly represent the population, and different samples may give different results.

Comparing Results

Making predictions is only one use of statistics. Suppose you have recently developed a new headache/pain remedy that you call Ache-Away. Should you produce Ache-Away in quantity and make it available to the public? That would depend on, among other factors, whether Ache-Away is more effective than the old remedy. How can you determine that?

One way might be to administer both remedies to two separate groups of people, collect data on the results, and then statistically analyze that data to determine if Ache-Away is more effective than the old remedy. And what if the results of this test showed Ache-Away to be more effective? How certain can you be that *this* particular test administration is indica- tests of these two remedies? Perhaps the group taking the old remedy (the control group) and the group taking Ache-Away (the treatment group) are so dissimilar that the results were due *not* to the pain remedies but to the differences between the groups.

It is possible that the results of this test differ greatly from the results that you would get if you tried the test several more times. You certainly do not want to foist a questionable drug upon an unsuspecting public based on atypical test results. How certain can you be that you should put your faith in the results of your tests? As you can see, the problems of comparing headache remedy results can produce headaches of their own!

This type of statistical procedure is discussed in chapters 6 and 7.

Probability

One of the most familiar uses of statistics is to determine the chance of some occurrence. For example, what are the chances that it will rain tomorrow or that the Chicago Cubs will win a World Series? These kinds of probabilities, although interesting, are not the variety

Coin tossing and card drawing are two such examples. A *fair coin* (one that is not weighted or *fixed*) has an equal chance of landing heads as landing tails. A standard deck of playing cards has 52 different cards—13 of each suit (hearts, clubs, diamonds, and spades)—and each card or suit has an equal chance of being drawn. This kind of event forms the basis of your understanding of probability and enables you to find solutions to everyday problems that seem far removed from coin tossing or card drawing.

Probability helps us describe our conclusions in a way that takes the uncertainty from random sampling into account. Probability is discussed in detail in Chapter 4.

Common Statistics Mistakes

Whether you are new to the study of statistics or you have been using statistical analysis for years, you likely will make (or already have made) errors in your application of principles explained in this book. In fact, certain errors are committed so frequently in statistical analysis that we feel obliged to call them to your attention in the hopes that, by being prepared for them, you will avoid their pitfalls. Appendix A lists these common mistakes.

Chapter Check-Out

True or False: Statistics is about things only in the past, so you cannot use it to make predictions.

True or False: An inference is an educated guess, made after analyz-

True or False: A population is a randomly chosen portion of a larger

True or False: Statistical analysis will allow you to be absolutely sure of your conclusions.

True or False: Probability helps to explain statistical conclusions when dealing with the uncertainties of particular events or situations.

True or False: It is not possible to describe an outcome with confidence when comparing two results.

Chapter 2

INTERPRETING GRAPHIC DISPLAYS

Chapter Check-In

Displaying categorical data in simple graphic formats such as bar charts and pie charts

Displaying quantitative variable data in simple graphic formats such as dot plots, frequency histograms, and stem-and-leaf plots

Using box plots to display numerical measures of data

Interpreting graphic displays to make conclusions about the distribution of the variable

Understanding scatter plots

Common Core Standard: Interpreting Categorical and Quantitative Data

Represent data with plots on the real number line—dot plots, histograms, and box plots (S.ID.1). Interpret differences in shape, center, and spread in the context of the data sets, accounting for possible effects of extreme data points, or outliers (S.ID.3).

raphic displays frequently appear in the Common Core Mathematics curriculum. You should be able to summarize, represent, and interpret data on a single count or measurement variable or on two categorical and quantitative variables and interpret linear models of graphic

Pie charts and bar charts are graphic displays of data for categorical variables. Dot plots, stem-and-leaf plots, **histograms,** and box-and-whisker plots are graphic displays of data for numerical variables. This chapter discusses how to construct these graphic displays.

As an example, consider the yearly expenditures of a college undergraduate. After collecting her data (expense records) for the past year, she organizes the expenditures, shown in Table 2-1.

Yearly Expenses of College Undergraduate

	Amount
	$30,000
	$15,300
	$5,500
Transportation, insurance (auto/medical), clothing, and miscellaneous	$5,200

These figures, although presented in categories, do not allow for easy analysis. The reader must expend extra effort in order to compare amounts spent or relate individual proportions to the total. For ease of analysis, these data can be presented pictorially.

Bar Charts

One way to pictorially display the numbers shown in Table 2-1 is with a The following vertical bar chart shows the expenditures of a college undergraduate for the past year.

Yearly Expenses
College Undergraduate

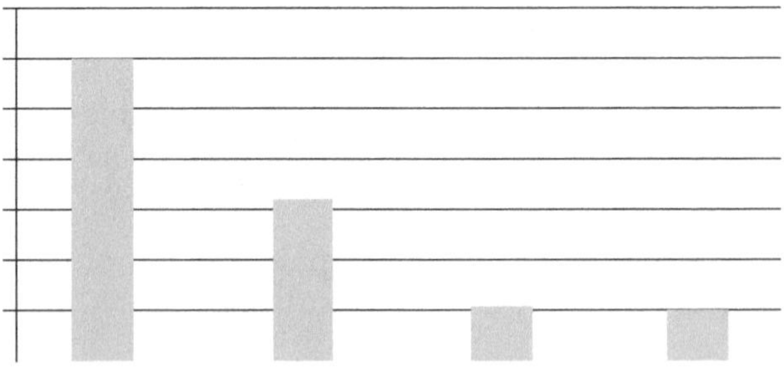

Comparing the size of the bars, you can quickly see that tuition fees are nearly double the room-and-board expenses and room-and-board fees are nearly triple the expenses for books and labs and miscellaneous (transportation, insurance, clothing, etc.).

A bar chart also may be placed on its side with the bars going horizontally, as shown in the following figure.

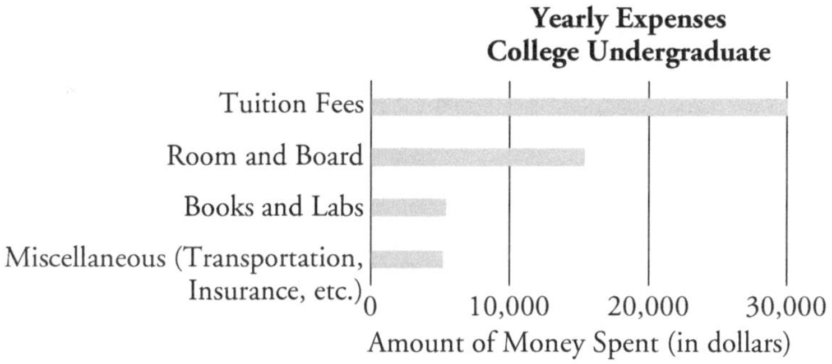

In each bar chart, vertical or horizontal, the amounts are ordered from highest to lowest or lowest to highest, making the chart clearer and easier to understand. Space is left between each of the bars in order to define the categories as being different.

The bottom line in the vertical bar chart and the left side in the horizontal bar chart indicate 0. Although typical, this presentation need not always be used. Finally, although the lengths of the bars may be different, their thicknesses are the same.

Pie Charts

Bar charts have a limitation: It is difficult to see what portion of the total each item comprises. If knowing about a part of the whole is an important consideration, a **pie chart** is a better choice for showing the same data. A pie chart also may display each category's percentage of the total. The data from the undergraduate expenditures is shown in the pie chart

The parts of the circle (or *pie*) match in size each category's percentage of

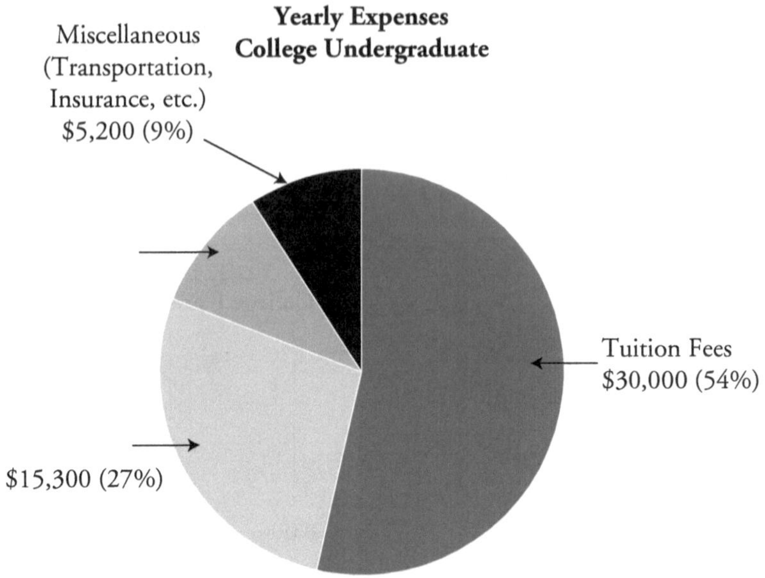

Dot Plots

are used for quantitative variables. Typically used for a small set of values, a dot plot uses a *dot* for each unit of measurement. For the preceding undergraduate expense data, if we just wanted to compare the dollar values and were not concerned with the categories, a dot plot would look like the following.

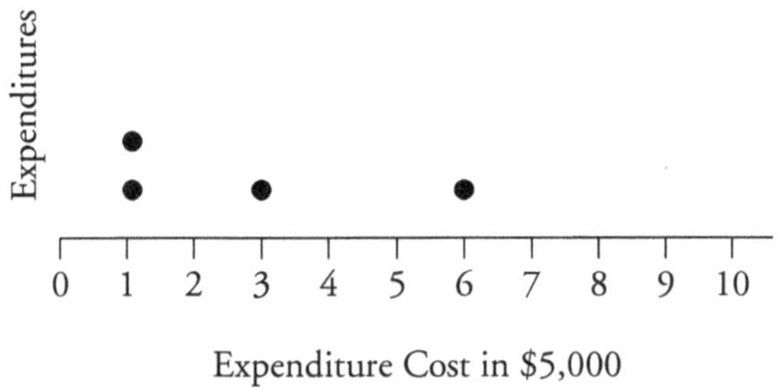

Data may be expressed using a single line. An **ogive** (a cumulative line

gradual slope. An ogive, however, is not the ideal graphic for showing comparisons between categories because it simply combines the values in each category, thus indicating an *accumulation* (an increasing or decreasing total). If you simply want to keep track of a total and your individual values are periodically combined, an ogive is an appropriate display.

For example, if you saved $300 in both January and April and $100 in each of February, March, May, and June, an ogive for accumulated savings for this 6-month period would look like this:

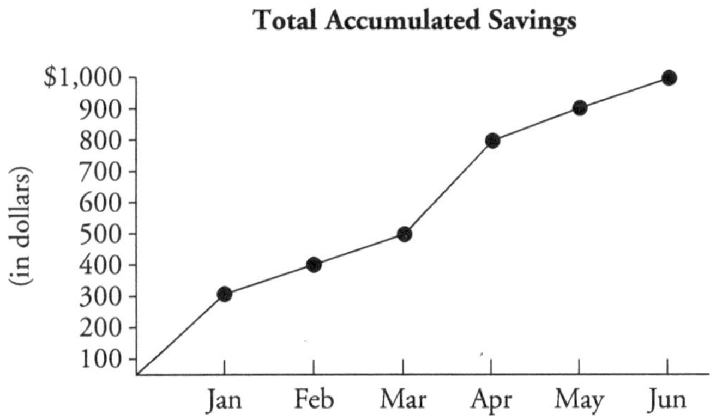

An ogive displays a running total. Although each individual month's avings could be expressed in a bar chart (as shown below), you could not easily see the amount of total growth or loss, as you can in an ogive.

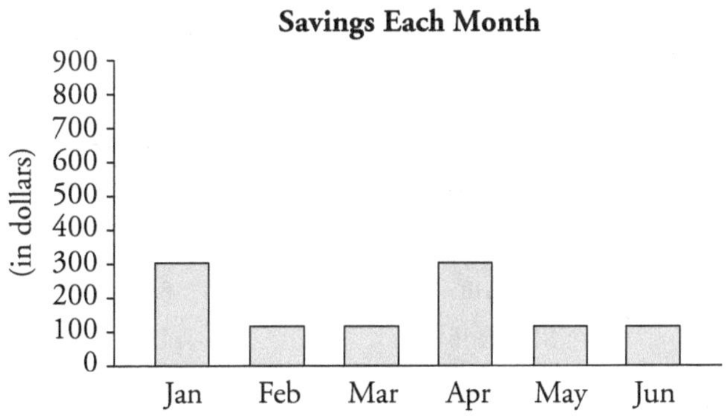

Frequency Histograms

One of the more commonly used graphic displays in statistics is the **frequency histogram,** which in some ways is similar to a bar chart; it tells how many items are in each numerical category. For example, suppose that after a garage sale, you want to determine which items were the most popular: the high-priced items, the low-priced items, and so forth. Let's say you sold a total of 32 items for the following prices: $1, $2, $2, $2, $5, $5, $5, $5, $7, $8, $10, $10, $10, $10, $11, $15, $15, $15, $19, $20, $21, $21, $25, $25, $29, $29, $29, $30, $30, $30, $35,

ranged in price from $1 to $35. First, divide this **range** of $1 to $35 into a number of categories, called **class intervals.** Typically, no fewer than 5 and no more than 20 class intervals work best for a frequency histogram.

Choose the first class interval to include your lowest (smallest value) data and make sure that no overlap exists so that one piece of data does not fall into two class intervals. For example, you would not have your first class interval be $1 to $5 and your second class interval be $5 to $10 because the four items that sold for $5 would belong in both the first and the second intervals. Instead, use $1 to $5 for the first interval and $6 to $10 for the second. Class intervals are mutually exclusive.

First, make a table of how your data is distributed (see Table 2-2). The number of observations that falls into each class interval is called the **class**

Note that each class interval has the same width. That is, $1 to $5 has a width of $5, inclusive; $6 to $10 has a width of $5, inclusive; $11 to $15

Distribution of Items Sold at Garage Sale

Class Interval	*Class Frequency*
$1 to $5	8
$6 to $10	6
$11 to $15	4
$16 to $20	2
$21 to $25	

has a width of \$5, inclusive; and so forth. From the data, a frequency histogram would look like this.

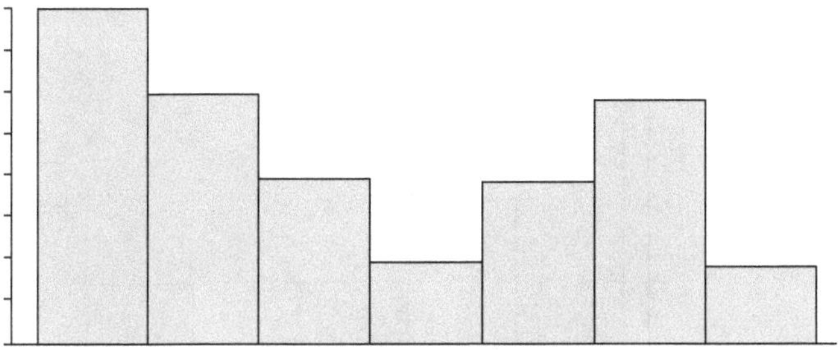

Unlike in a bar chart, the class intervals are drawn immediately adjacent to each other.

Relative frequency histograms

Relative frequency is the ratio of class frequency to total number of measures. A **relative frequency histogram** is a graphic display that looks like a bar chart with columns that shows the proportion (percent) of the data values of a particular interval. It uses the same information as a frequency histogram but compares each class interval to the total number of items. For example, the first interval (\$1 to \$5) contains 8 of the total 32 items, so the relative frequency of the first class interval is $\frac{8}{32}$ or $\frac{1}{4}$ or 0.25. Calculate the relative frequency by dividing the frequency by its total number (see Table 2-3).

Distribution of Items Sold at Garage Sale, Including Relative Frequencies

Class Interval	*Class Frequency*	*Relative Frequency*
\$1 to \$5	8	0.25
\$6 to \$10	6	0.1875
\$11 to \$15	4	0.125
\$16 to \$20	2	0.0625
\$21 to \$25		

The only difference between a frequency histogram and a relative frequency histogram is that the vertical axis uses relative or proportional frequency, as shown in the following figure, instead of simple class

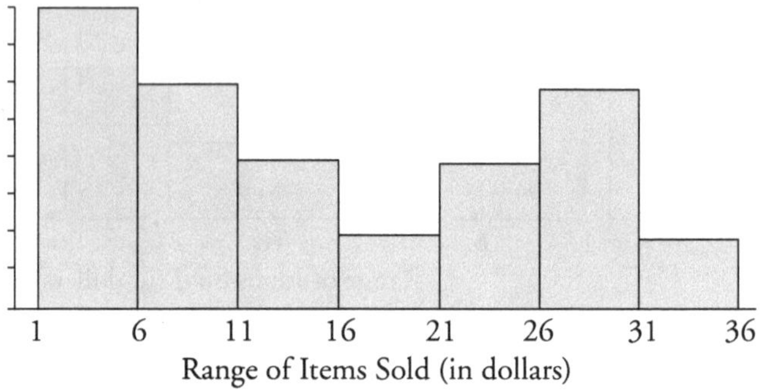

Frequency polygons

Relative frequencies of class intervals also can be shown in a **frequency**

In this type of chart, the frequency of each class is indicated by points or dots drawn at the midpoints of each class interval. Those points are then connected by straight lines.

Comparing the frequency polygon (shown below) to the frequency histogram on p. 19, you see that the major difference is that points replace

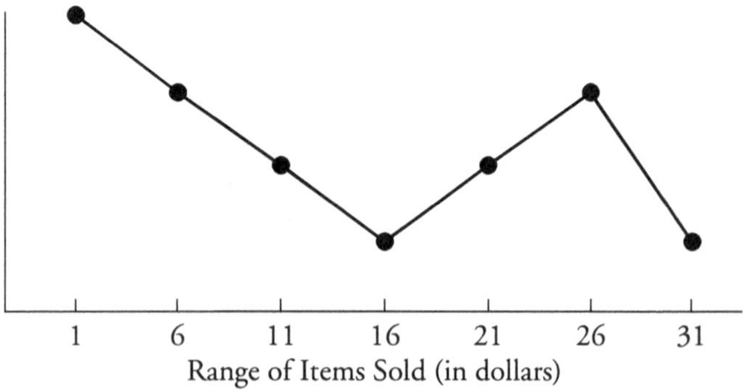

Gender (male or female), types of automobile owned (sedan, sports car, pickup truck, van, and so forth), and religious affiliations (Christian, Jewish, Muslim, and so forth) are all qualitative data. On the other hand, **quantitative data** can be measured in amounts: age in years, annual salaries, inches of rainfall. Typically, qualitative data are better displayed in bar charts, and quantitative data are better displayed in histograms.

Frequency distributions

Frequency distributions are similar to frequency polygons (see p. 20); however, instead of straight lines, a frequency distribution uses a smooth curve to connect the points and, similar to a graph, is plotted on two axes. The horizontal axis from left to right (or *x*-axis) indicates the different possible values of some variable (a phenomenon where observations vary from trial to trial). The vertical axis from bottom to top (or *y*-axis) measures frequency or how many times a particular value occurs.

For example, the following **bell-shaped curve** figure is the most common distribution. It is often called a **normal distribution.** The *x*-axis might indicate annual income (the values would be in thousands of dol- axis might indicate frequency (millions of people or percentage of working population). Notice that the highest percentage of the working population would, therefore, have an annual income in the middle of the dollar values. The lowest percentages would be at the extremes of the values: nearly 0 and extremely high.

Bell-Shaped Curve

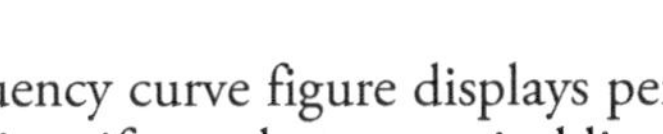

Notice that this frequency curve figure displays perfect **symmetry.** In a symmetrical distribution, if you draw a vertical line through the middle, one half (the left side) is the mirror image of the other half (the right side). Not all frequency curves are perfectly symmetrical. If a curve is not symmetrical, it is *skewed.* Skewed curves can be positive or negative (see pp. 36–37 for a description of symmetric and skewed distributions in

This asymmetrical distribution is probably a more accurate representation of the annual income of working Americans than a symmetrical distribution of a bell-shaped curve.

Positively Skewed Curve

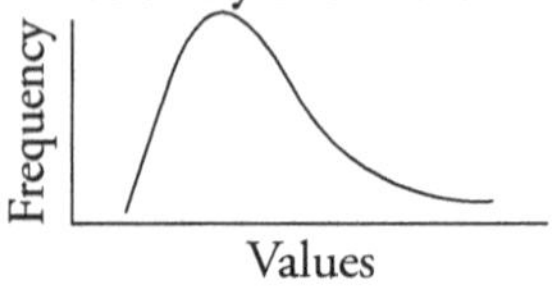

negatively skewed curve figure shown below is *skewed to the left.* This means that the bulk of the values tend to pile up near the right of the graph and taper off toward the left (called the *tail* of the distribution).

Negatively Skewed Curve

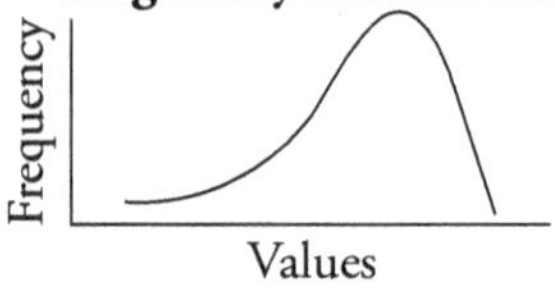

The next frequency curve figure shows a **J-shaped curve.** It is sometimes *exponential curve* because it displays an asymmetrical skewed distribution—an extreme case of a negative skew.

J-Shaped Curve

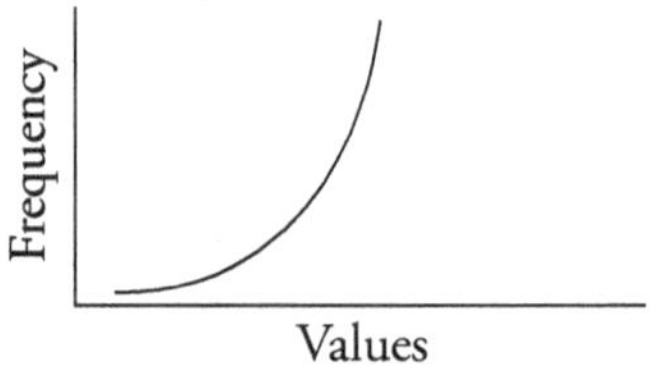

Unlike the symmetrical bell curve, a **bimodal curve** (p. 23) is double-peaked and has two high points (two modes). While a distribution has only one mean and one median, it can have more than one mode. For example, the distribution curve of the annual income of working Ameri

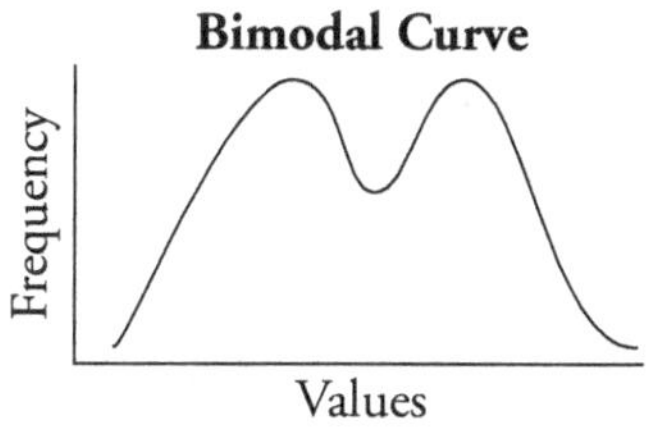

Stem-and-Leaf Plots

Another useful graphic display is the **stem-and-leaf plot.** It is similar to a histogram in that it shows the range of data, where the data are concentrated, if there are any **outliers** (occasional extremely high or extremely low values), and the general shape of the distribution.

For example, look at the following data—test scores of 17 high school students: 69, 75, 77, 79, 82, 84, 87, 88, 89, 89, 89, 90, 91, 93, 96, 100, and 100. The stem-and-leaf plot takes all but the last digit of each score as the stem and uses the remaining digit as the leaf.

As an example, for the score of 69, 6 is the stem and 9 is the leaf; for the next three grades (75, 77, and 79), 7 is the stem, and 5, 7, and 9 are the

Note, too, that along the extreme left side of the following stem-and-leaf plot is a vertical column that keeps a running count or total. (Some stem-and-leaf plots do not include this running total.) Having a running total enables the reader to quickly locate the median. (Median is discussed further in Chapter 3.)

The completed stem-and-leaf plot for the high school students' test scores looks like Table 2-4.

Stem-and-Leaf Plot of Students' Test Scores

Running Count	*Stem*	*Leaves*
	6	9
	7	5, 7, 9
	8	2, 4, 7, 8, 9, 9, 9

Notice that, as in a histogram, each stem determines a class interval and, also as in a histogram, the class intervals are all equal. (In this case, the number of digits that can be used range from a possible low of 0 to a possible high of 9; e.g., 60 to 69.) All 17 scores are displayed in the stem-and-leaf plot so that you can see not only the frequencies and the shape of the distribution, but also the actual value of every score.

Box Plots (Box-and-Whisker Plots)

sometimes called **box-and-whisker plots,** take the stem-and-leaf plot one step further. A box plot displays a number of values of a distribution of numbers:

The median value

The lower quartile (Q_1)

The upper quartile (Q_3)

interquartile range (**IQR**), the difference between the lower and upper quartiles

The symmetry of the distribution

The highest and lowest values

Use the set of values in Table 2-5 to examine each of the preceding items.

Verbal Scores of 20 Students

440	490	520	540	560	560	580	580
630	650	660	680	710	730	740	740

(the middle value in a set that has been ordered lowest to highest) is the value above which half of the remaining values fall and below which the other half of the remaining values fall. Because there is an even number of scores in our example (20), the median score is the average of the two middle scores (10th and 11th)—580 and 600—

lower quartile (Q or 25th percentile) is the median of the bottom

sixth scores—520 and 540—or 530. Therefore, the lower-quartile score

upper quartile (Q_3 or 75th percentile) is the median score of the top half. The top half of this set consists of the last ten numbers: 600, 610, 630, 650, 660, 680, 710, 730, 740, and 740. The median of these ten is again the average of the fifth and sixth scores—in this case, 660 and 680—or 670. So 670 is the upper-quartile score for this set of 20 numbers.

A box plot can now be constructed as follows: The left side of the box indicates the lower quartile, the right side of the box indicates the upper quartile, and the line inside the box indicates the median. A horizontal line is then drawn from the lowest value of the distribution through the box to the highest value of the distribution. (This horizontal line is the "whiskers.")

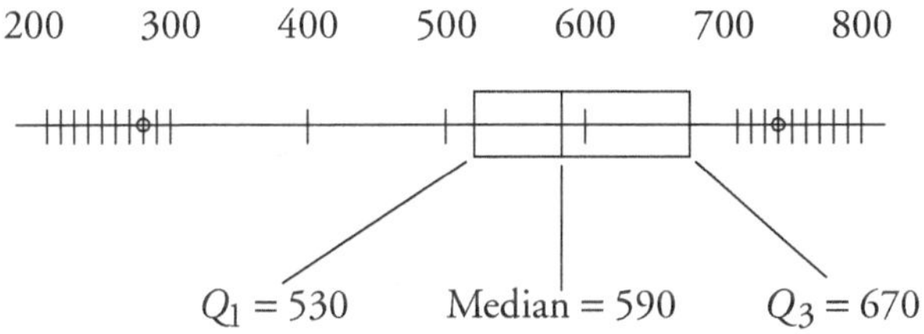

Without reading the actual values, you can see by looking at the box plot that the scores range from a low of 280 to a high of 740; that the lower
Q_1) is at 530; that the median is at 590; and that the upper
Q_3) is at 670. Because the median is slightly nearer the lower quartile than the upper quartile and the interquartile range is situated far to the right of the range of values, the distribution departs from

Scatter Plots

Sometimes you want to display information about the relationship involving two different phenomena. These data would be called **bivariate** or paired data. A **scatter plot** is the best way to see correlations. A **cor-**
describes the statistical relationships between two variables. For

Law School Students' Prep Times and Test

	1	**2**	**3**	**4**	**5**	**6**	**7**	**8**
	7	9	5	1	8	4	3	6
	23	25	14	5	22	15	11	17

One dot would then be plotted for each examinee, giving a total of only eight dots, yet displaying 16 pieces of numerical information. For example, Students 1 studied for 7 days and received a score of 23. Student 1's dot would be plotted at a vertical of 23 and a horizontal of 7, as shown in the following figure.

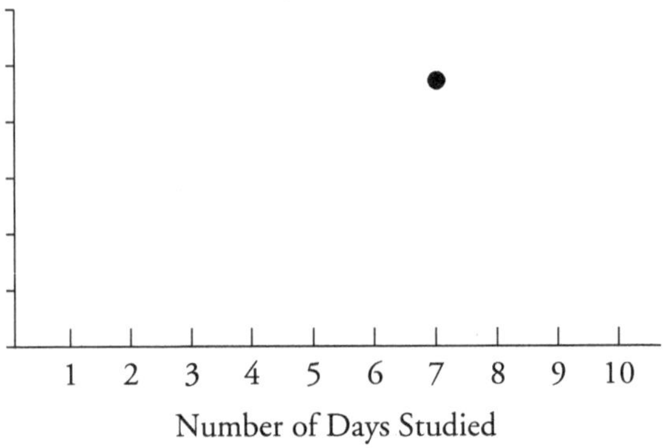

A completed scatter plot would look like the following.

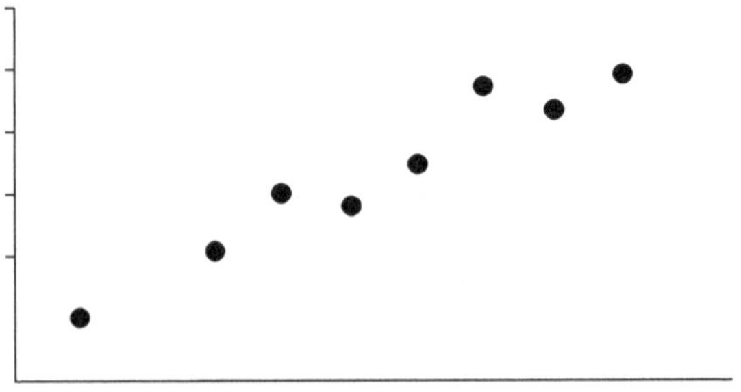

There is a strong **positive relationship** between the number of days studied and the score on the exam; that is, the data displayed indicates that an increase in days studied for the exam correlates with an increase in score achieved. A **negative relationship** would be indicated if the dots suggested a line going down from left to right, meaning that as one variable increases, the other decreases. Mathematically, when the slope of the function is positive, we say that there is a *positive relationship*. If the slope is negative, we say that there is a *negative relationship*. In our example, we are working with the function that describes the relationship between "days studied" and "score earned." Can you see an imaginary line with positive slope in the scatter plot above? Therefore, it is concluded that a positive relationship exists.

No relationship would be indicated if the scatter plot dots suggested a completely horizontal line, a completely vertical line, or no line at all. The following scatter plots display no relationship between the variables

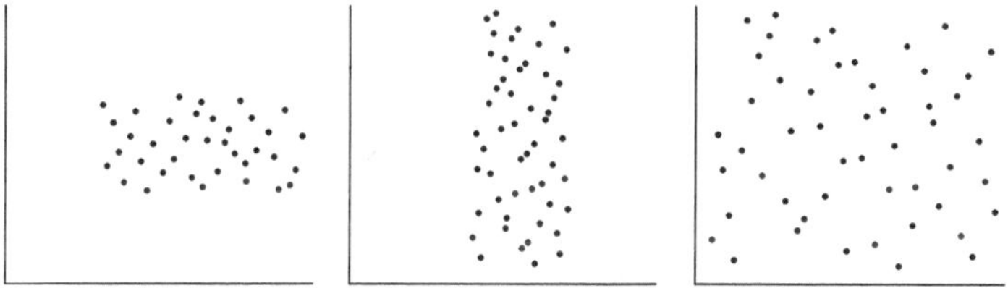

These relationships are discussed in more detail in Chapter 8.

Chapter Check-Out

Questions

A used car salesman takes inventory and finds that he has a total of 125 cars to sell. Of these, 97 are the 2016 model, 11 are the 2015 model, 12 are the 2014 model, and 5 are the 2013 model. Which two types of graphs are most appropriate to display the data? Construct one of the graphs.

Given the bar graph shown here, take the information and construct a scatter plot and answer the following questions.

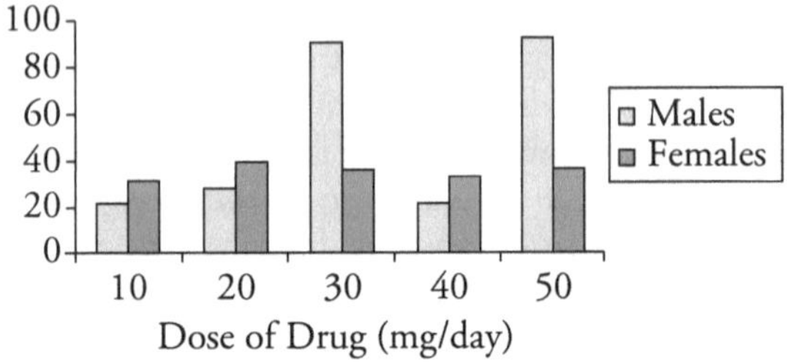

Which drug dose (mg per day) shows the least difference in the increase of hair growth between males and females?
Does increasing the drug dose (mg per day) necessarily result in increasing hair growth in females?
Does increasing the drug dose (mg per day) have the same effect on hair growth for males and females?

Using this ogive shown in the graph, answer the following questions.

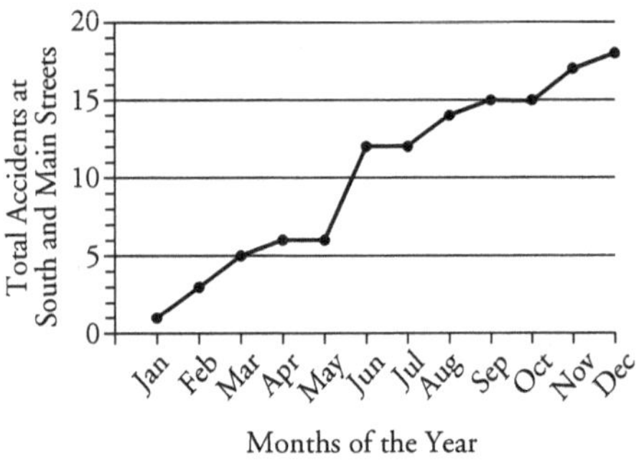

How many total accidents have occurred at the intersection of South and Main streets over one year?
How many occurred between June and August?
What interval saw the greatest increase in the rate of accidents?

Pie chart or bar chart

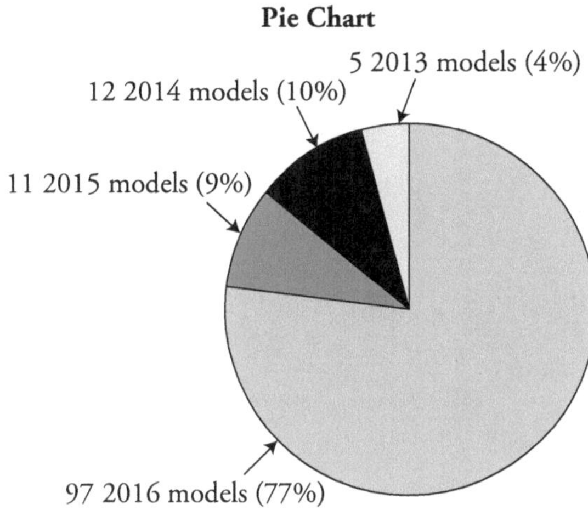

10 mg/day; **(b)** No; **(c)** No

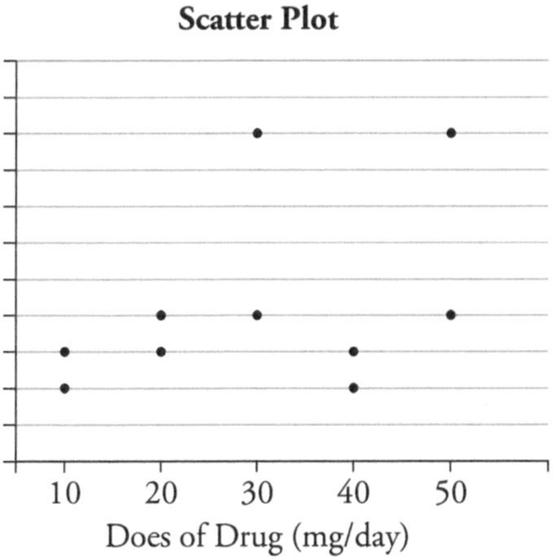

18; **(b)** 2; **(c)** May through June

Chapter 3

MEASURES OF CENTRAL TENDENCY AND VARIABILITY

Chapter Check-In

- ❑ Calculating and understanding the central tendencies of data
- ❑ Using measures of variability to describe the spread in data
- ❑ Discussing the principal types of measurement scales

Common Core Standard: Interpreting Categorical and Quantitative Data

Use statistics appropriate to the shape of the data distribution to compare center (median, mean) and spread (interquartile range, standard deviation) of two or more data sets (S.ID.2). Interpret differences in shape, center, and spread in context of the data sets, accounting for possible effects of extreme data points, or outliers (S.ID.3). Use the mean and standard deviation of a data set to fit it to a normal distribution (S.ID.4). Construct and interpret two-way frequency tables of data when two categories are associated with each object being classified (S.CP.4).

Chapter 2 discussed how to graph numerical information—to transform it into a picture—to get a better sense of the data. Another way to get a sense of data is to use numerical measures, certain numbers that give special insight into your values. Two types of numerical measures are important in statistics: **measures of central tendency** and **measures of variation.** Each of these individual measures can provide information about the entire set of data. The study of measures is a way to evaluate research experiments emphasized in Common Core Mathematics.

Measures of Central Tendency

Measures of central tendency are numbers that tend to cluster around the "middle" of a set of values. Four such middle numbers are the mean, the median, the midrange, and the mode.

For example, suppose your earnings for the past week were the values shown in Table 3-1.

Table 3-1 Earnings for the Past Week

Day	*Amount*
Monday	$1,350
Tuesday	$1,150
Wednesday	$1,100
Thursday	$1,350
Friday	$150

Mean

You could express your daily earnings from Table 3-1 in a number of ways. One way is to use the average, or **mean,** of the data set. The arithmetic mean is the sum of the measures in the set divided by the number of measures in the set. Totaling all the measures and dividing by the number of measures, you get $\$5{,}100 \div 5 = \$1{,}020$.

Median

Another measure of central tendency is the **median,** which is defined as the middle value when the numbers are arranged in increasing or decreasing order. When you order the daily earnings shown in Table 3-1, you get $150, $1,100, $1,150, $1,350, and $1,350. The middle value is $1,150; therefore, $1,150 is the median.

If there is an even number of items in a set, the median is the average of the two middle values. For example, if we had four values—4, 10, 12, and 26—the median would be the average of the two middle values, 10 and 12; in this case, 11 is the median. The median may sometimes be a better indicator of central tendency than the mean, especially when there are outliers, or extreme values.

Example 1: Given the four annual salaries of a corporation shown in Table 3-2, determine the mean and the median.

Table 3-2 Four Annual Salaries

Position	*Salary*
CEO	$1,000,000
Manager	$50,000
Administrative Assistant	$30,000
Custodian	$20,000

The mean of these four salaries is $275,000. The median is the average of the middle two salaries, or $40,000. In this instance, the median appears to be a better indicator of central tendency because the CEO's salary is an extreme outlier, causing the mean to lie far from the other three salaries.

Midrange

The mean of the minimum and the maximum of the value in a data set is called the **midrange,** where the minimum is the lowest value in a data set and the maximum is the highest value in a data set. In the set of weekly earnings in Table 3-2, the midrange is the mean of $20,000 and $1,000,000, which is $510,000.

Mode

Another indicator of central tendency is the **mode,** or the value that occurs most often in a set of numbers. In the set of weekly earnings in Table 3-1, the mode would be $1,350 because it appears twice and the other values appear only once.

Notation and formulas

The mean of a sample is typically denoted as $\bar{x}$ (read as *x* bar). The mean of a population is typically denoted as μ (pronounced *mew*). The sum (or total) of measures is typically denoted with a Σ. The formula for a sample mean is

$$\bar{x} = \frac{\Sigma x}{n} = \frac{x_1 + x_x + \ldots + x_n}{n}$$

where n is the number of values.

Mean for grouped data

Occasionally, you may have data that consist not of actual values but rather of **grouped measures.** For example, you may know that in a certain working population, 32 percent earn between $25,000 and $29,999;

40 percent earn between \$30,000 and \$34,999; 27 percent earn between \$35,000 and \$39,999; and the remaining 1 percent earn between \$80,000 and \$85,000. This type of information is similar to that presented in a frequency table. (Refer to Chapter 2 for information about frequency tables and graphic displays.) Although you do not have precise individual measures, you still can compute measures for **grouped data,** data presented in a frequency table.

The formula for a sample mean for grouped data is

$$\bar{x} = \frac{\sum fx}{n}$$

where x is the midrange of the interval, f is the frequency for the interval, fx is the product of the midpoint times the frequency, and n is the number of values.

For example, if 8 is the midrange of a class interval and there are 10 measurements in the interval, $fx = 10(8) = 80$, the sum of the 10 measurements in the interval.

Σfx denotes the sum of all the products in all class intervals. Dividing that sum by the number of measurements yields the sample mean for grouped data.

For example, consider the information shown in Table 3-3.

Table 3-3 Distribution of the Prices of Items Sold at a Garage Sale

Class Interval	*Frequency (f)*	*Midpoint (x)*	*fx*
\$1.00 to \$5.99	8	3.495	27.96
\$6.00 to \$10.99	6	8.495	50.97
\$11.00 to \$15.99	4	13.495	53.98
\$16.00 to \$20.99	2	18.495	36.99
\$21.00 to \$25.99	4	23.495	93.98
\$26.00 to \$30.99	6	28.495	170.97
\$31.00 to \$35.99	2	33.495	66.99
	$n = 32$		$\Sigma fx = 501.84$

Substituting into the formula:

$$\bar{x} = \frac{\sum fx}{n} = \frac{501.84}{32} = 15.6825$$

Therefore, the average price of items sold was about \$15.68. The value may not be the exact mean for the data because the actual values are not always known for grouped data.

Median for grouped data

As with the mean, the median for grouped data may not necessarily be computed precisely, because the actual values of the measurements may not be known. In that case, you can find the particular interval that contains the median and then approximate the median.

Using Table 3-3, you can see that there are a total of 32 measures. The median is between the 16th and 17th measure; therefore, the median falls within the \$11.00 to \$15.99 interval. The formula for the best approximation of the median for grouped data is

$$\text{median} = L + \frac{w}{f_{\text{med}}}\left(0.5n - \Sigma f_b\right)$$

where L is the lower class limit of the interval that contains the median, n is the total number of measurements, w is the class width, f_{med} is the frequency of the class containing the median, and Σf_b is the sum of the frequencies for all classes before the median class.

Consider the information in Table 3-4.

Table 3-4 Distribution of Prices of Items Sold at a Garage Sale

Class Interval	*Frequency (f)*
\$1.00 to \$5.99	8
\$6.00 to \$10.99	6
\$11.00 to \$15.99	4
\$16.00 to \$20.99	2
\$21.00 to \$25.99	4
\$26.00 to \$30.99	6
\$31.00 to \$35.99	2
	n = 32

As we already know, the median is located in class interval \$11.00 to \$15.99. So $L = 11$, $n = 32$, $w = 4.99$, $f_{\text{med}} = 4$, and $\Sigma f_b = 14$.

Substituting into the formula:

$$
\begin{aligned}
\text{median} &= L+\frac{w}{f_{\text{med}}}\left(0.5n-\Sigma f_b\right)\\
&=11+\frac{4.99}{4}\left(0.5(32)-14\right)\\
&=11+\frac{4.99}{4}(16-14)\\
&=11+\frac{4.99}{4}(2)\\
&=11+2.495\\
&=13.495
\end{aligned}
$$

Rounding, the median price is \$13.50.

Symmetric distribution

In a distribution displaying perfect symmetry (**symmetric distribution**), the mean, the median, and the mode are all at the same point, as shown in the following figure.

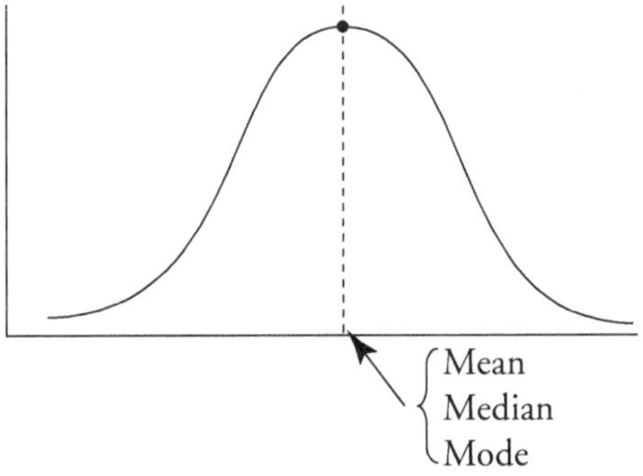

Skewed curves

As you have seen, an outlier can significantly alter the mean of a series of numbers, whereas the median will remain at the center of the series. In such a case, the resulting curve drawn from the values will appear to be **skewed,** tailing off rapidly to the left or right. In the case of negatively skewed or positively skewed curves, the median remains in the center of these three measures. In a negatively skewed curve, such as the following, mean < median < mode.

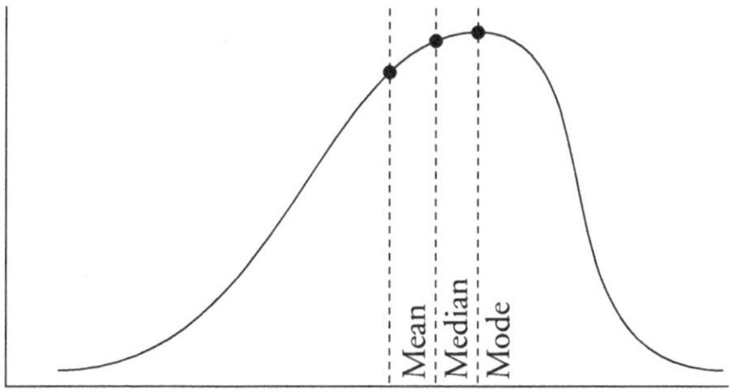

In a positively skewed curve, such as the following, mode < median < mean.

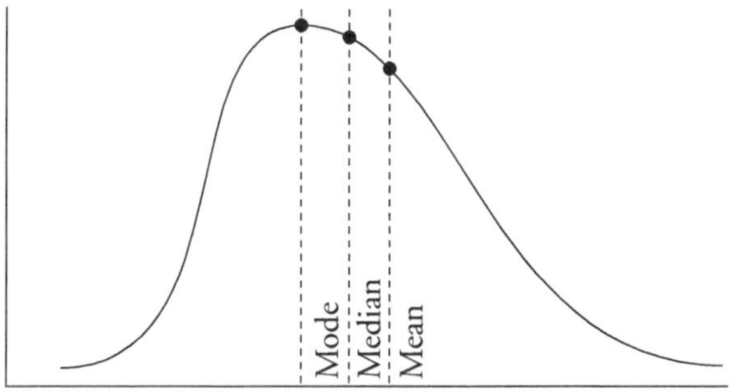

Measures of Variability

Measures of central tendency locate only the center of a distribution of measures. Other measures often are needed to describe data. For example, consider the two sets of numbers presented in Table 3-5.

Table 3-5 Earnings of Two Employees

Daily Earnings of Employee A	*Daily Earnings of Employee B*
\$200	\$200
\$210	\$20
\$190	\$400
\$201	\$0
\$199	\$390
\$195	\$10
\$205	\$200
\$200	\$380

The mean, the median, and the mode of each employee's daily earnings all equal \$200. Yet, there is significant difference between the two sets of numbers. For example, the daily earnings of Employee A are much more consistent than those of Employee B, which show great variation. This example illustrates the need for **measures of variation** or **spread.**

Range

The most elementary measure of variation is **range.** Range is defined as the difference between the largest and smallest values. The range is a single number. To say that the range is from \$190 to \$200, although informative, is not really a correct use of the term. The range for Employee A is \$210 − \$190 = \$20; the range for Employee B is \$400 − \$0 = \$400.

Deviation and variance

The **deviation** is defined as the distance of the measurements away from the mean. In Table 3-5, Employee A's earnings have considerably less deviation than Employee B's do. The **variance** is defined as the sum of the squared deviations of n measurements from their mean divided by $(n - 1)$.

So, from Table 3-5, the mean for Employee A is \$200, and the deviations from the mean are as follows:

$$0, +10, -10, +1, -1, -5, +5, 0$$

The squared deviations from the mean, therefore, are the following:

$$0, 100, 100, 1, 1, 25, 25, 0$$

The sum of these squared deviations from the mean equals 252. Dividing by $(n - 1)$, or $8 - 1$, yields $\frac{252}{7} = 36$. So, the variance is 36.

For Employee B, the mean is also \$200, and the deviations from the mean are as follows:

$$0, -180, +200, -200, +190, -190, 0, +180$$

The squared deviations, therefore, are the following:

$$0, 32{,}400, 40{,}000, 40{,}000, 36{,}100, 36{,}100, 0, 32{,}400$$

The sum of these squared deviations equals 217,000. Dividing by $(n - 1)$ yields $\frac{217,000}{7} = 31,000$. So, the variance is 31,000.

Although the two employees earned the same total amounts, there is significant difference in variance between their daily earnings.

Standard deviation

The **standard deviation** is defined as the positive square root of the variance; thus, the standard deviation of Employee A's daily earnings is the positive square root of 36, or 6. The standard deviation of Employee B's daily earnings is the positive square root of 31,000, or about 176.

Notation

s^2 denotes the variance of a sample.

σ^2 denotes the variance of a population.

s denotes the standard deviation of a sample.

σ denotes the standard deviation of a population.

Empirical rule: The normal curve

One practical significance of the standard deviation is that, with mound-shaped (bell-shaped) distributions, the following rules apply:

- The interval from one standard deviation below the mean to one standard deviation above the mean contains approximately 68 percent of the measurements.

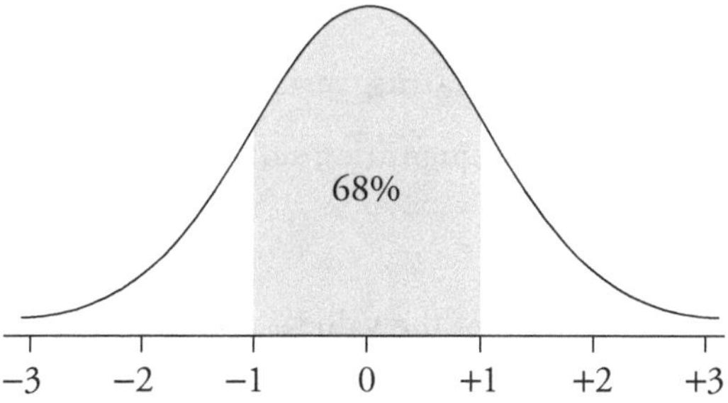

- The interval from two standard deviations below the mean to two standard deviations above the mean contains approximately 95 percent of the measurements.

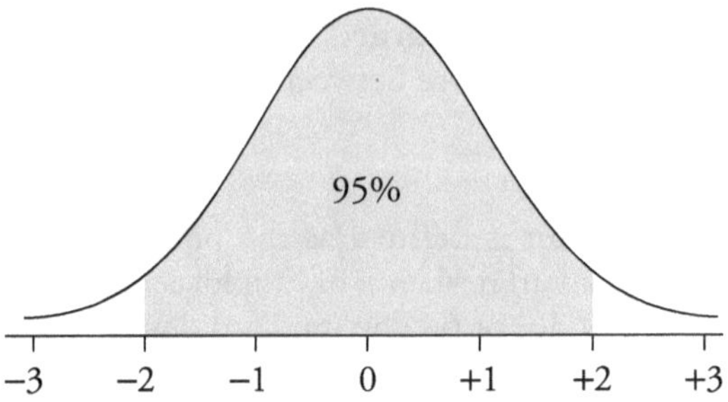

- The interval from three standard deviations below the mean to three standard deviations above the mean contains nearly all the measurements.

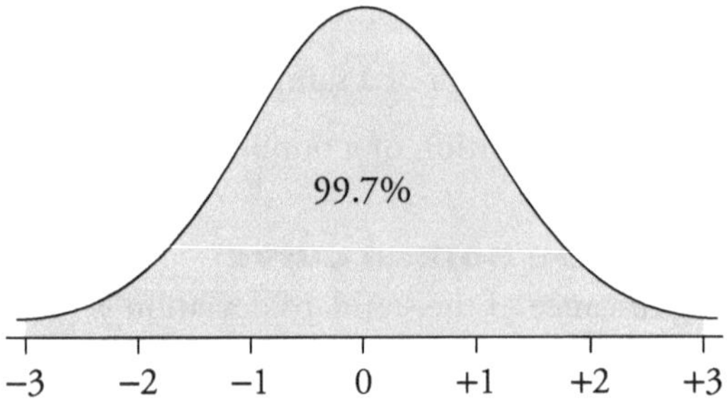

These **mound-shaped curves** usually are called **normal distributions** or normal curves.

Shortcut formulas

A shortcut method of calculating variance and standard deviation, $\sum x^2 - \frac{(\sum x)^2}{n}$, requires two quantities: sum of the values and sum of the squares of the values.

$$\Sigma x = \text{sum of the values}$$

$$\Sigma x^2 = \text{sum of the squares of the values}$$

For example, using these six values 3, 9, 1, 2, 5, and 4:

$$\begin{aligned}\sum x &= 3+9+1+2+5+4 = 24 \\ \sum x^2 &= 3^2+9^2+1^2+2^2+5^2+4^2 \\ &= 9+81+1+4+25+16 \\ &= 136\end{aligned}$$

The quantities are then substituted into the shortcut formula to find $\sum(x-\overline{x})^2$.

$$\begin{aligned}\sum(x-\overline{x})^2 &= \sum x^2 - \frac{(\sum x)^2}{n} \\ &= 136 - \frac{(24)^2}{6} \\ &= 136 - \frac{576}{6} \\ &= 40\end{aligned}$$

The variance and standard deviation are now found as before (see pp. 38–39):

$$s^2 = \frac{\sum(x-\overline{x})^2}{n-1} = \frac{40}{5} = 8$$

$$s = \sqrt{s^2} = \sqrt{8} = 2.828$$

Sensitivity to outliers

An **outlier** is a data value that has an abnormal distance from other values in a data set. In a distribution, it is best to choose a measure of center, or a measure of variation, that is less sensitive to outliers. To illustrate outliers' influence on a measure of center, consider the number of patients visiting a dermatologist's office in a day over a 12-day period:

4, 18, 20, 21, 21, 21, 23, 23, 24, 25, 25, 43

The outliers are 4 and 43. The mean, 22.33, takes outliers into account, but if the outliers are not taken into account, the mean is 22.1. Note that the median is not sensitive to outliers.

Now let's see how outliers can influence measures of variations. As previously defined, the range is the difference between the maximum and minimum number of patients. Therefore, in this example, if outliers are taken into account, the range is 39. If we calculate the range without considering outliers, the range is 7. As you can see, the range is extremely sensitive to outliers.

Now let's check the standard deviation for the data set of the number of patients. The standard deviation is 8.58, including outliers. Without outliers, the standard deviation is 2.28. Therefore, the standard deviation is sensitive to outliers. In comparison with the range, we may say that that standard deviation is less sensitive to outliers.

Percentile

The *N*th **percentile** is defined as the value such that *N* percent of the values lie below it. So, for example, a score of 5 percent from the top score would be the 95th percentile because it is above 95 percent of the other scores.

|← 95% of Scores →|

0 0 0 1 1 1 1 1 1 2 2 2 2 2 3 3 3 4 4 5 5 5 6 6 6 6 7 7 8 9 9 9 10 11 11 11 12 12 (13) 15

95th Percentile

Quartiles and interquartile range

The **lower quartile** (Q_1) is defined as the 25th percentile; thus, 75 percent of the measures are above the lower quartile. The **middle quartile** (Q_2) is defined as the 50th percentile, which is, in fact, the median of all the measures. The **upper quartile** (Q_3) is defined as the 75th percentile; thus, only 25 percent of the measures are above the upper quartile.

The **interquartile range (IQR)** is the value for $Q_3 - Q_1$. Like the range, it is a single value. The following figure illustrates the locations of the median and the quartiles for a set of 20 test scores.

Scores on Test

17 23 28 38 40 41 41 45 47 60 61 61 63 80 80 81 82 85 88 95

Q_1(40.5) Median Q_2(60.5) Q_3(80.5)

|← Interquartile Range (IQR) (40) →|

Measurement Scales

Different measurement scales allow for different levels of exactness, depending upon the characteristics of the variables being measured. The four types of scales available in statistical analysis are as follows:

- **Nominal scale:** A scale that measures data by name only. For example, religious affiliation (measured as Christian, Jewish, Muslim, and so forth); political affiliation (measured as Democratic, Republican, Libertarian, and so forth); or style of automobile (measured as sedan, sports car, SUV, and so forth).
- **Ordinal scale:** A scale that measures by rank order only. Other than rough order, no precise measurement is possible. For example, medical condition (measured as satisfactory, fair, poor, guarded, serious, or critical); socioeconomic status (measured as lower class, lower-middle class, middle class, upper-middle class, upper class); or military officer rank (measured as lieutenant, captain, major, lieutenant colonel, colonel, general). Such rankings are not absolute but rather relative to each other: Major is higher than captain, but we cannot measure the exact difference in numerical terms. Is the difference between major and captain equal to the difference between colonel and general? We cannot say.
- **Interval scale:** A scale that measures by using equal intervals. Here you can compare differences between pairs of values. The Fahrenheit temperature scale, measured in degrees, is an interval scale, as is the centigrade scale. The temperature difference between 50°C and 60°C (10 degrees) equals the temperature difference between 80°C and 90°C (10 degrees).
- **Ratio scale:** Similar to an interval scale, a ratio scale includes a 0 measurement that signifies the point at which the characteristic being measured vanishes (absolute zero). For example, income (measured in dollars, with 0 equal to no income at all), years of formal education, items sold, and so forth are all ratio scales.

Chapter Check-Out

Questions

1. A carabiner machine can make up to 200 carabiners per hour.
   (a) For 4 consecutive hours, the carabiner machine makes 120, 118, 124, and 112 carabiners, respectively. Find the average rate of carabiner production.
   (b) During the fifth hour, the machine breaks after making only 21 carabiners. The remainder of the hour is spent repairing it. Normally, the machine must average 115 carabiners per hour, over the course of an 8-hour day, in order to meet its quota. What must the average rate be for the next 3 hours to meet this quota?

2. A meteorologist measures the average rate of rainfall in order to test the saying "When it rains, it pours." Her data, showing the rainfall rate over the course of a rainy morning, are as follows:

Time of Day	*Rainfall Rate (inches/minute)*
10:00 a.m. to 10:15 a.m.	0.011
10:15 a.m. to 10:30 a.m.	0.007
10:30 a.m. to 10:45 a.m.	0.001
10:45 a.m. to 11:00 a.m.	0.010
11:00 a.m. to 11:15 a.m.	0.003
11:15 a.m. to 11:30 a.m.	0.045
11:30 a.m. to 11:45 a.m.	0.002
11:45 a.m. to 12:00 p.m.	0.006

   (a) Find the median and mean.
   (b) Calculate the variance and standard deviation.

3. Can a distribution have the mean equal to the median, but the mode not equal to either mean or median? If so, give an example.

Answers

1. (a) 118.5; (b) 141.67
2. (a) median = 0.0065, mean = 0.0106;
   (b) variance: $s^2 = 0.0002$, standard deviation: $s = 0.014$
3. Yes; a symmetric but bimodal distribution is a good example.

Chapter 4
PROBABILITY

Chapter Check-In

- ❑ Applying classic probability theory to simple events
- ❑ Analyzing combinations of simple events based on these rules
- ❑ Analyzing combinations of independent and dependent events
- ❑ Analyzing the probability of joint occurrences
- ❑ Evaluating mutually exclusive outcomes and those that are not mutually exclusive
- ❑ Learning about conditional probability and Bayes' Theorem
- ❑ Learning about probability distributions, especially the binomial distribution

Common Core Standard: Conditional Probability and the Rules of Probability

Understand independence and conditional probability and use them to interpret data (S.CP). Use the rules of probability to compute probabilities of compound events in a uniform probability model (S.CP.6-8). Use probability to make decisions (S.MD).

Probability theory plays a central role in statistics and in Common Core Mathematics. After all, statistical analysis is applied to a collection of data in order to discover something about the underlying events. These events may be complementary to one another—for example, mutually exclusive—but the individual choices involved are assumed to be random. Alternatively, we may sample a population at random and make inferences about the population as a whole from the sample by using statistical analysis. Therefore, a solid understanding of probability theory—the study of random events—is necessary to understand how statistical analysis works and also to correctly interpret the results.

You may have an intuition about probability. Pierre Laplace, a great French mathematician, said probability theory is nothing but "common sense reduced to calculation." As you will see, in some cases, probability theory seems obvious. But be careful: Occasionally, a seemingly obvious answer will turn out to be wrong—because sometimes your intuition about probability will fail. Even in seemingly simple cases, it is best to follow the rules of probability as described in this chapter rather than rely on your hunches.

Two best-known models in the formalization of probability are from the mathematicians Andrey Kolmogorov and Sir David Cox. Kolmogorov's formulation of events is associated with a collection of sets, and Cox's formulation of events is a derivation of a certain set of probability axioms. In both models the laws of probability are the same.

Classic Theory

The classic theory of probability underlies much of probability in statistics. Briefly, this theory states that the chance of a particular outcome occurring is determined by the ratio of the number of favorable outcomes (or "successes") to the total number of outcomes. Expressed as a formula,

$$P(A) = \frac{\text{number of favorable outcomes}}{\text{total number of possible outcomes}}$$

For example, the probability of randomly drawing an ace from a well-shuffled standard deck of 52 playing cards is equal to the ratio $\frac{4}{52}$. Four is the number of favorable outcomes (the number of aces in the deck), and 52 is the number of total outcomes (the number of cards in the deck). The probability of randomly selecting an ace in one draw from a deck of cards is, therefore, $\frac{4}{52}$, or 0.077. In statistical analysis, probability is usually expressed as a decimal and ranges from a low of 0 (no chance) to a high of 1 (certainty).

The classic theory assumes that all outcomes have an equal likelihood of occurring. In the example just cited, each card must have an equal chance of being chosen—no card is in any way more likely to be chosen than any other card.

The classic theory pertains only to outcomes that are **mutually exclusive** (or a **disjoint occurrence**), which means that those outcomes may not occur at the same time. For example, one coin flip can result in a head or a tail, but one coin flip cannot result in a head *and* a tail. So, the outcome of a head and the outcome of a tail are said to be mutually exclusive in one coin flip, as is the outcome of an ace and a king in one card being drawn.

Relative Frequency Theory

The **relative frequency theory of probability** holds that if an experiment is repeated an extremely large number of times and a particular outcome occurs a percentage of the time, then the probability of that outcome is close to that particular percentage.

For example, if a machine produces 10,000 widgets one at a time and 1,000 of those widgets are faulty, the probability of that machine producing a faulty widget is approximately 1,000 out of 10,000, or 0.10.

Probability of Simple Events

Example 1:

(a) What is the probability of simultaneously flipping three coins—a penny, a nickel, and a dime—and having all three land heads?

Using the classic theory, determine the ratio of the number of favorable outcomes to the number of total outcomes. Table 4-1 lists all possible outcomes.

Table 4-1 Possible Outcomes of Penny, Nickel, and Dime Flipping

Outcome	*Penny*	*Nickel*	*Dime*
1	Head	Head	Head
2	Head	Head	Tail
3	Head	Tail	Head
4	Head	Tail	Tail
5	Tail	Head	Head
6	Tail	Head	Tail
7	Tail	Tail	Head
8	Tail	Tail	Tail

There are eight different outcomes, only one of which is favorable (Outcome 1: all three coins landing heads); therefore, the probability of three coins landing heads is $\frac{1}{8}$, or 0.125.

(b) What is the probability of exactly two of the three coins landing heads?

Again, there are the eight total outcomes, but in this case only three favorable outcomes (outcomes 2, 3, and 5); thus, the probability of exactly two of three coins landing heads is $\frac{3}{8}$, or 0.375.

Independent Events

Each of the three coins being flipped in the preceding example is what is known as an independent event. **Independent events** are defined as outcomes that are not affected by other outcomes. In other words, the flip of the penny does not affect the flip of the nickel, and vice versa.

Dependent Events

Dependent events, on the other hand, are outcomes that are affected by other outcomes. Consider the following example.

Example 2: What is the probability of randomly drawing an ace from a standard deck of 52 playing cards and then drawing an ace again from the same deck of cards, without returning the first drawn card back to the deck?

For the first draw, the probability of a favorable outcome is $\frac{4}{52}$, as explained earlier; however, after that first card has been drawn, the total number of outcomes is now 51 because a card has been removed from the deck. And if that first card drawn resulted in a favorable outcome (an ace), there would now be only three aces in the deck. If that first card drawn were *not* an ace, the number of favorable outcomes would remain at four. So, the second draw is a dependent event because its probability changes depending upon what happens on the first draw.

If, however, you replace that drawn card back into the deck and shuffle well again before the second draw, then the probability for a favorable outcome for each draw will be equal $\left(\frac{4}{52}\right)$, and these events will be independent.

Probability of Joint Occurrences

Another way to compute the probability of all three flipped coins landing heads is as a series of three different events: First flip the penny, then flip the nickel, and then flip the dime. Will the probability of landing three heads still be 0.125? To answer this question, you'll need to use the multiplication rule.

Multiplication rule

To compute the probability of **joint occurrence** (two or more independent events all occurring), multiply their probabilities.

For example, the probability of the penny landing heads is $\frac{1}{2}$, or 0.5; the probability of the nickel landing heads is $\frac{1}{2}$, or 0.5; and the probability of the dime landing heads is $\frac{1}{2}$, or 0.5. Multiplying these together, $0.5 \times 0.5 \times 0.5 = 0.125$, which is what you determined with the classic theory by assessing the ratio of the number of favorable outcomes to the number of total outcomes. The notation for joint occurrence is

$$P(A \cap B) = P(A) \times P(B)$$

which is read: "The probability of *A* and *B* both happening is equal to the probability of *A* times the probability of *B*."

Using the **multiplication rule,** you also can determine the probability of drawing two aces in a row from a standard deck of 52 playing cards. The only way to draw two aces in a row from a deck of cards is for both draws to be favorable. For the first draw, the probability of a favorable outcome is $\frac{4}{52}$. But because the first draw is favorable, only three aces are left among 51 cards. So, the probability of a favorable outcome on the second draw is $\frac{3}{51}$. For both events to happen, you simply multiply those two probabilities together:

$$\frac{4}{52} \times \frac{3}{51} = \frac{12}{2{,}652} = 0.0045$$

Note that these probabilities are not independent. If, however, you had decided to return the initial card drawn back to the deck before the second draw, then the probability of drawing an ace on each draw is $\frac{4}{52}$, because these events are now independent. Drawing an ace twice in a row, with the odds being $\frac{4}{52}$ both times, gives the following:

$$\frac{4}{52} \times \frac{4}{52} = \frac{16}{2,704} = 0.0059$$

In either case, you use the multiplication rule because you are computing probability for favorable outcomes in all events.

Addition rule

Given mutually exclusive events, finding the probability of *at least one* of them occurring is accomplished by adding their probabilities. This is known as the **addition rule.**

Example 3: What is the probability of one coin flip resulting in at least one head or at least one tail?

The probability of one coin flip landing heads is 0.5, and the probability of one coin flip landing tails is 0.5. Are these two outcomes mutually exclusive in one coin flip? Yes, they are. You cannot have a coin land both heads and tails in one coin flip; therefore, you can determine the probability of at least one head or one tail resulting from one flip by adding the two probabilities:

$$0.5 + 0.5 = 1 \text{ (or certainty)}$$

Example 4: What is the probability of at least one spade or one club being randomly chosen in one draw from a standard deck of 52 playing cards? The probability of drawing a spade in one draw is $\frac{13}{52}$; the probability of drawing a club in one draw is $\frac{13}{52}$. These two outcomes are mutually exclusive in one draw because you cannot draw both a spade and a club in one draw; therefore, you can use the addition rule to determine the probability of drawing at least one spade or one club in one draw:

$$\frac{13}{52} + \frac{13}{52} = \frac{26}{52} = 0.50$$

Outcomes That Are Not Mutually Exclusive

For the addition rule to apply, the events must be mutually exclusive. Now consider the following example.

Example 5: What is the probability of the outcome of at least one head in two coin flips?

Should you add the two probabilities as in the preceding examples? In Example 3, you added the probability of getting a head and the probability of getting a tail because those two events were mutually exclusive in one flip. In Example 4, the probability of getting a spade was added to the probability of getting a club because those two outcomes were mutually exclusive in one draw. Now when you have two flips, should you add the probability of getting a head on the first flip to the probability of getting a head on the second flip? Are these two events mutually exclusive?

Of course, they are not mutually exclusive. You can get an outcome of a head on one flip and a head on the second flip. So, because they are not mutually exclusive, you cannot use the addition rule. If you did use the addition rule, you would get

$$\frac{1}{2} + \frac{1}{2} = \frac{2}{2} = 1$$

or certainty, which is absurd. There is no certainty of getting at least one head on two flips. (Try it several times, and see that there is a possibility of getting two tails and no heads.)

To find the answer to the question in Example 5, "What is the probability of the outcome of at least one head in two coin flips?" keep reading.

Double-Counting

By using the addition rule in a situation that is not mutually exclusive, you are **double-counting.** One way of realizing that you are double-counting is to use the classic theory of probability to solve the question in

Example 5: List all the different outcomes when flipping a coin twice and assess the ratio of favorable outcomes to total outcomes (see Table 4-2).

Table 4-2 All Possible Outcomes of Flipping the Same Coin Twice

First Flip	*with*	*Second Flip*
Head	+	Head
Head	+	Tail
Tail	+	Head
Tail	+	Tail

There are four total outcomes. Three of the outcomes have at least one head; therefore, the probability of throwing at least one head in two flips is $\frac{3}{4}$, or 0.75, not 1. But if you had used the addition rule, you would have added the two heads from the first flip to the two heads from the second flip and gotten four heads in four flips, $\frac{4}{4} = 1$. But the two heads in that first pair constitute only one outcome; so, by counting both heads for that outcome, you are double-counting because this is the joint-occurrence outcome that is not mutually exclusive.

To use the addition rule in a situation that is not mutually exclusive, you must subtract any events that double-count. In this case:

$$\frac{1}{2} + \frac{1}{2} - \frac{1}{4} = \frac{3}{4} = 0.75$$

The notation, therefore, for at least one favorable occurrence in two events is

$$P(A \cup B) = P(A) + P(B) - P(A \cap B)$$

which is read: "The probability of at least one of the events *A* or *B* equals the probability of *A* plus the probability of *B* minus the probability of their joint occurrence." (Note that if they are mutually exclusive, then $P(A \cap B)$—the joint occurrence—equals 0, and you simply add the two probabilities.)

Example 6: What is the probability of drawing either a spade or an ace from a standard deck of 52 playing cards?

The probability of drawing a spade is $\frac{13}{52}$; the probability of drawing an ace is $\frac{4}{52}$. But the probability of their joint occurrence (an ace of spades) is $\frac{1}{52}$. Thus,

$$\begin{aligned} P(A \cup B) &= P(A) + P(B) - P(A \cap B) \\ &= \frac{13}{52} + \frac{4}{52} - \frac{1}{52} \\ &= \frac{16}{52} \\ &= \frac{4}{13} \\ &= 0.3077 \end{aligned}$$

Conditional Probability

Common Core Mathematics emphasizes your understanding of using independent and conditional probability to interpret data. **Conditional probability** is the probability of an event (A), given that another event (B) has already occurred. Sometimes you have more information than simply total outcomes and favorable outcomes; hence, you are able to make more informed judgments regarding probabilities.

Example 7: In a Mayan village, there are 60 women and 40 men. Twenty of those women are 70 years of age or older; five of the men are 70 years of age or older (see Table 4-3).

Table 4-3 Distribution of People in a Mayan Village

	70 or Older	*69 or Younger*	*Total*
Women	20	40	60
Men	5	35	40
Total	25	75	100

(a) What is the probability that a person selected at random in that village will be a woman?

Because women constitute 60 percent of the total population, the probability is 0.60.

(b) What is the probability that a person 70 years of age or older selected at random in that village will be a woman?

This question is different because the probability of A (being a woman) given B (the person in question being 70 years of age or older) is now conditional upon B (being 70 years of age or older). Because women number 20 out of the 25 people in the 70-or-older group, the probability of this latter question is $\frac{20}{25}$, or 0.80.

Conditional probability is found using this formula for dependent events:

$$P(A \mid B) = \frac{P(A \cap B)}{P(B)}$$

This formula is read: "The probability of A given B equals the probability of A and B divided by the probability of B." The vertical bar in the expression $A|B$ is read *given that* or *given.* Note that A must depend on B so that we can use the above formula to find the probability of A given B.

Since event B has already occurred, we take B as our sample space. Hence, the sample space is reduced. Probability was defined in the beginning of this chapter as the "ratio of the number of favorable outcomes over the number of all possible outcomes." Therefore, the probability of A and B is divided by the probability of B.

Bayes' Theorem

Bayes' theorem describes the relationship between conditional probability and its reverse form using this formula:

$$P(A \mid B) = \frac{P(B \mid A) \cdot P(A)}{P(B)}$$

For example, assume there is a 30 percent chance that Sarah will get a grade of A on her first exam. Conditional on getting a grade of A on the first exam is a chance of getting a grade of B on the second exam, or a 20 percent chance. Also, there is a 90 percent chance that she will get a grade of B on the second exam. With the help of the above formula, we can calculate the probability that Sarah will get a grade of A on the first exam if we know that she will receive a grade of B on the second exam.

$$P(\text{A first exam} \mid \text{B second exam})$$
$$= \frac{P(\text{B second exam} \mid \text{A first exam}) \cdot P(\text{A first exam})}{P(\text{B second exam})}$$
$$= \frac{(30\% \cdot 20\%)}{90\%} \approx 0.07$$

Probability Distributions

A **probability distribution** is a graphic display of the probability—$P(x)$—for any value of x. Consider the number of possible outcomes of two coins being flipped (see Table 4-4). Table 4-5 shows the probability distribution of the results of flipping two coins.

Table 4-4 Possible Outcomes of Two Flipped Coins

Head + head	=	2 heads
Head + tail	=	1 head
Tail + head	=	1 head
Tail + tail	=	0 heads

Table 4-5 Probability Distribution: Number of Heads

x(Number of Heads)	*P(x)*
0	$\frac{1}{4}$, or 0.25
1	$\frac{1}{2}$, or 0.50
2	$\frac{1}{4}$, or 0.25

The following figure displays this information graphically.

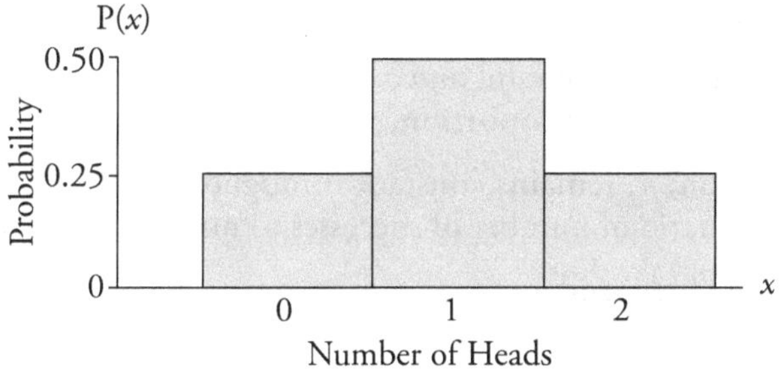

Discrete versus continuous variables

The number of heads resulting from coin flips can be counted only in integers (whole numbers). The number of aces drawn from a deck can be counted only in integers. These "countable" numbers are known as **discrete variables:** 0, 1, 2, 3, 4, and so forth. No value between two variables is possible. For example, 2.6 is not possible.

However, qualities such as height, weight, temperature, and distance can be measured using fractions or decimals as well: 34.27, 102.26, and so forth. These are known as **continuous variables.**

Total of probabilities of discrete variables

The probability associated with a discrete variable lies somewhere between 0 and 1, inclusive. For example, the probability of tossing one head in two coin flips is, as earlier, 0.50. The probability of tossing two heads in two coin flips is 0.25, and the probability of tossing no heads in two coin flips is 0.25. The sum (total) of probabilities for all values of x always equals 1. For example, note that in Table 4-5, adding the probabilities for all the possible outcomes yields a sum of 1.

The Binomial

A discrete variable that can result in only one of two outcomes is called a **binomial.** For example, a coin flip is a binomial variable, but drawing a card from a standard deck of 52 playing cards is not. Whether a drug is successful or unsuccessful in producing results is a binomial variable, as is whether a machine produces perfect or imperfect widgets.

Binomial experiments

Binomial experiments require the following elements:

- The experiment consists of a number of identical events (n).
- Each event has only one of two mutually exclusive outcomes. (These outcomes are called successes and failures.)
- The probability of a successful outcome is equal to some percentage, which is identified as a **proportion,** π.
- This proportion, π, remains constant throughout all events and is defined as the ratio of number of successes to number of trials.
- The events are independent.

- Given all these elements, the binomial formula can be applied (x = number of favorable outcomes; n = number of events):

$$P(x) = \frac{n!}{x!(n-x)!}\pi^x(1-\pi)^{n-x}$$
$$n! = n(n-1)(n-2)\ldots(3)(2)(1)$$

Example 8: A coin is flipped ten times. What is the probability of getting exactly five heads?

Using the binomial formula, where n (the number of events) is given as 10, x (the number of favorable outcomes) is given as 5, and the probability of landing a head in one flip is 0.5:

$$\begin{aligned} P(x) &= \frac{n!}{x!(n-x)!}\pi^x(1-\pi)^{n-x} \\ &= \frac{10!}{5!5!}(0.5^5)(1-0.5)^5 \\ &= \frac{10\cdot9\cdot8\cdot7\cdot6\cdot5\cdot4\cdot3\cdot2\cdot1}{5\cdot4\cdot3\cdot2\cdot1\cdot5\cdot4\cdot3\cdot2\cdot1}(0.03125)(0.03125) \\ &= 252(0.03125)(0.03125) \\ &= 0.246 \end{aligned}$$

So, the probability of getting exactly five heads in ten flips is 0.246, or approximately 25 percent.

Binomial table

Because probabilities of binomial variables are so common in statistics, tables are used to alleviate having to continually use the formula. Refer to Table 1 in Appendix B, and you will find that given $n = 10$, $x = 5$, and $\pi = 0.5$, the probability is 0.2461.

Mean and standard deviation

The mean of the **binomial probability distribution** is determined by the following formula:

$$\mu = n\pi$$

where π is the proportion of favorable outcomes and n is the number of events.

The standard deviation of the binomial probability distribution is determined by this formula:

$$\sigma = \sqrt{n\pi(1-\pi)}$$

Example 9: What is the mean and standard deviation for a binomial probability distribution for ten coin flips of a fair coin? Hint: A fair coin is a coin that is not weighted to produce a predetermined outcome (to guarantee either a head or tail).

Because the proportion of favorable outcomes of a fair coin falling heads (or tails) is $\pi = 0.5$, simply substitute into the formulas:

$$\mu = n\pi = 10(0.5) = 5$$

$$\sigma = \sqrt{n\pi(1-\pi)} = \sqrt{10(0.5)(0.5)} = \sqrt{2.5} = 1.581$$

The probability distribution for the number of favorable outcomes is shown in the following figure.

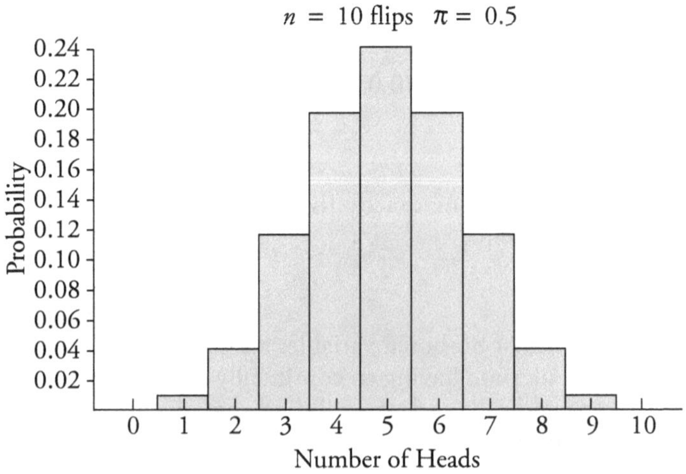

Note that this distribution appears to display symmetry. Only a binomial distribution with $\pi = 0.5$ will be truly symmetric. All other binomial distributions will be skewed. Binomial probability distributions and use of the normal approximation to replace the probability histogram are discussed in Chapter 5.

Chapter Check-Out

Questions

1. A dartboard is divided into 20 equal wedges, ignoring the bull's-eye.
   (a) If only 6 of the 20 wedges are worth points, what is the probability of scoring on one throw (assuming you can hit the board)? Give your answer in decimal form.
   (b) What is the chance of hitting at least one scoring region in three consecutive throws? Give your answer in decimal form.
2. During a lecture, you begin repeatedly flipping an old, worn quarter. Amazingly, the first ten flips are all heads.
   (a) What is the chance of this, if the coin is fair? Give your answer in decimal form.
   (b) Suppose that the coin is not fair. If the probability of heads is 75 percent rather than 50 percent, what is the chance of your result? Give your answer in decimal form.
3. A woman has three sons. What is the probability that her next child will be a girl? Give your answer in decimal form.
4. Consider an ordinary (six-sided) die, assumed to be fair.
   (a) Use the binomial distribution to determine the probability that, in ten rolls of the die, you will see exactly four sixes. Give your answer in decimal form.
   (b) If you roll the die 100 times, what is the chance of rolling exactly 40 sixes? Give your answer in decimal form.
5. What is the probability that a student will correctly answer 7 out of 10 multiple-choice questions on a quiz by using blind guessing? (Each question has four answer choices.) Give your answer as a percent.
6. Two equally qualified candidates, Mr. X and Mr. Y, are competing in the Republican presidential primary. What is the probability that Mr. X will win 5 out of the 12 remaining states? Give your answer as a percent.

Answers

1. (a) 0.3; (b) 0.657
2. (a) 0.000977; (b) 0.0563
3. 0.5
4. (a) 0.0543; (b) 0.0000000182
5. 0.3%
6. 19%

Chapter 5
RANDOM SAMPLING

Chapter Check-In

- ❑ Examining the differences between populations and samples
- ❑ Learning about sampling distributions, sampling errors, and the central limit theorem
- ❑ Discussing in detail the properties of the normal distribution

Common Core Standard: Making Inferences and Justifying Conclusions

Understand statistics as a process for making inferences about population based on a random sample from that population (S.IC.1). Determine the sample size required for a desired margin of error. Make inferences and justify conclusions from sample surveys, experiments, and observational studies (S.IC.3-4). Define a random variable for a quantity of interest by assigning a numerical value to each event in a sample space (S.MD).

As discussed in Chapter 1, the field of inferential statistics enables you to make educated guesses about the numerical characteristics of large groups. The logic of sampling gives you a way to test conclusions about such groups using only a small portion of its members. Data are collected from selected elements of a population to represent a subset of the population.

In research statistics, it is almost impossible to study every single subject (or agent) of an experiment due to the time, expense, and resource limitations. Therefore, we will teach you how to select a small group of a given population, called a **sample size,** to make inferences about the entire population or group. When conducting hypothesis testing, the key is to select the right elements in the sampling to obtain reliable and valid research results. Of course, researchers generalize results from the sample size to the population with some degree of error. It is important to learn how to calculate the error and mention it in the conclusion.

In Common Core Mathematics, you should be able to make inferences and justify conclusions from sample surveys and experiments. This requires a deep understanding of statistics as the process for making inferences about population parameters based on a random sample from that population.

Populations, Samples, Parameters, and Statistics

A **population** is a group of phenomena that have something in common. The term often refers to a category of people or things, as in the following examples:

- All registered voters in Baltimore County
- Members of the Teamsters Union
- Americans who drink coffee at least once a week
- The record of the daily maximum temperatures in July for major U.S. cities
- The number of accidents that occurred in 2015 in the five largest U.S. cities due to drunk driving

Often, researchers want to obtain information about populations but do not have data for every person or thing in the population. For example, if a company's customer service division wanted to learn whether its customers were satisfied, it would not be practical (or perhaps even possible) to contact every individual who had purchased a product. Instead, the company might select a **sample** of the population. A sample is a smaller group of members of a population selected to represent the population. In order to use statistics to learn things about the population, the sample must be a random sample. A **random sample** is one in which every member of a population has an equal chance of being selected.

The most commonly used kind of sample is a simple random sample. It requires that every *possible* sample of the selected size has an equal chance of being used. But what does it mean that every sample has an equal chance of being used?

Consider this real-world scenario for conducting random sampling. A classroom teacher has five free tickets for an event at the science museum, but there are 30 students in the class. How can the teacher pick the five students who will receive the tickets using an unbiased selection method? The teacher can conduct a fair game by first assigning every student a number from 1 through 30, without duplicating numbers. Then the teacher can generate random numbers either electronically (using MATLAB or other software) or from a table of random numbers, 1 through 30. In a random sample, the first five matching numbers drawn will be identified as winners, and each of the corresponding students will receive a ticket to visit the science museum. By using this method, every student has an equal chance of being selected to receive a ticket.

In theory, random sampling should work each time. In practice, however, the problem is that the elements that are selected as a sample may not be as *random* as the population. This means that the sample elements may be more similar to one another than those of the population. Remember, a sample is intended to be representative of the entire population, so we need to be careful.

For example, consider different polls of voters for a presidential election. Is it best to choose all of the registered voters from the entire population of *one state* when selecting a reliable sample? Or is it better to select a *designated* number of registered voters from *each state* without considering the population of each state? To help answer these questions, researchers identify the characteristics of a population by selecting population *parameters*.

A **parameter** is a characteristic of a population. A **statistic** is a characteristic of a sample. Inferential statistics enables you to make an educated guess about a population parameter based on a statistic derived from a sample randomly drawn from that population. The following figure illustrates these relationships.

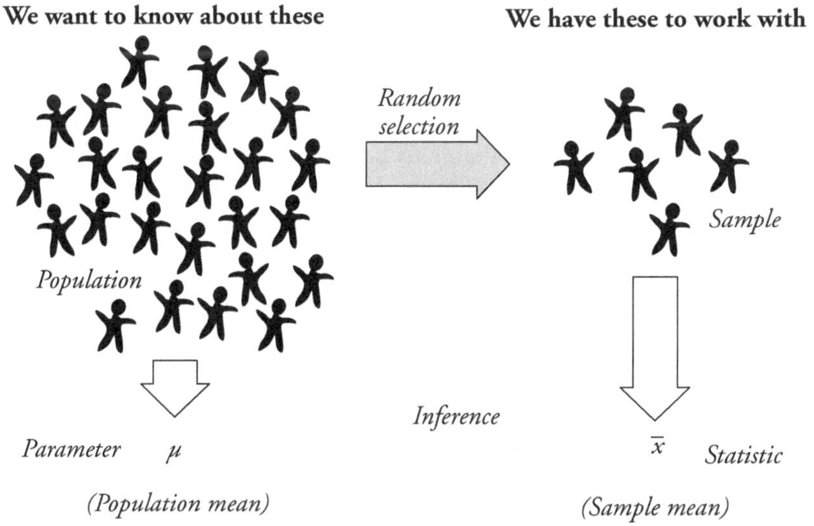

For example, say you want to know the mean income of the subscribers to a particular magazine—a parameter of a population. You draw a random sample of 100 subscribers and determine that their mean income is $27,500 (a statistic). You conclude that the population mean income is likely to be close to $27,500 as well. This example is one of statistical inference.

Different symbols are used to denote statistics and parameters, as listed in Table 5-1.

Table 5-1 Comparison of Sample Statistics and Population Parameters

	Sample Statistic	*Population Parameter*
Mean	$\overline{x}$	μ
Standard deviation	s	σ
Variance	s^2	σ^2

Sampling Distributions

Continuing with the earlier example of magazine subscribers, suppose that ten different samples of 100 people were drawn from the population, instead of just one sample. You would not expect the income means of these ten samples to be exactly the same because of **sampling variability** (the tendency of the same statistic derived from a number of random samples drawn from the same population to differ).

Suppose that the first sample of 100 magazine subscribers was "returned" to the population (made available to be selected again), another sample of 100 subscribers was selected at random, and the mean income of the new sample was derived. If this process were repeated ten times, it might yield the following sample means:

$27,500	$27,192	$28,736	$26,454	$28,527
$28,407	$27,592	$27,684	$28,827	$27,809

These ten values are part of a **sampling distribution.** The sampling distribution of a statistic (in this case, of a mean) is the **distribution** obtained by computing the statistic for all possible samples of a specific size drawn from the same population.

You can estimate the mean of this sampling distribution by adding the ten sample means and dividing by ten, which gives a distribution mean of $27,873 (rounded). Suppose that the mean income of the entire population of subscribers to the magazine is $28,000. (You usually do not know what it is.) You can see in the following figure that the first sample mean ($27,500) was not a bad estimate of the population mean, and that the mean of the distribution of ten sample means ($27,873) was even better. Estimation of the population mean becomes progressively more accurate as more samples are taken.

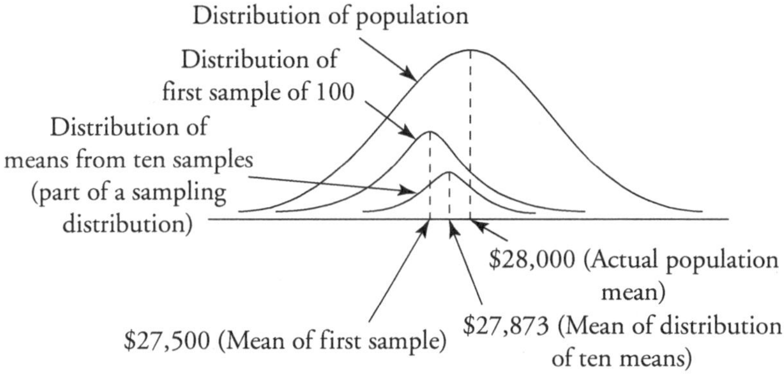

Random and Systematic Error

Two potential sources of error occur in statistical estimation—two reasons a statistic might misrepresent a parameter. **Random error** occurs as a result of sampling variability. The ten sample means in the magazine subscribers example on p. 64 differed from the true population mean because of random error. Some were below the true value; some were above it. Similarly, the mean of the distribution of the ten sample means was slightly lower than the true population mean. If ten more samples of 100 subscribers were drawn, the mean of that distribution—that is, the mean of those means—might be higher than the population mean.

Systematic error or **bias** refers to the tendency to consistently underestimate or overestimate a true value. Suppose that your list of magazine subscribers was obtained through a database of information about air travelers. The samples that you would draw from such a list would likely overestimate the population mean of all subscribers' income because lower-income subscribers are less likely to travel by air and many of them would be unavailable to be selected for the samples. This example would be one of bias.

In the following figure, both of the dot plots on the right illustrate systematic error (high bias). The results from the samples for these two situations do not have a center close to the true population value. Both of the dot plots on the left have centers close to the true population value (low bias).

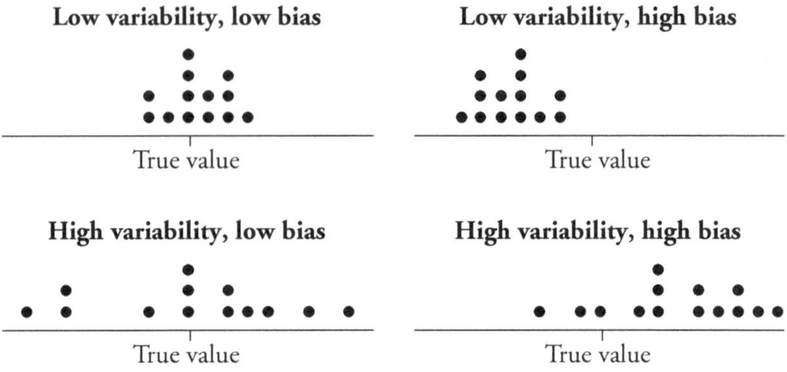

Central Limit Theorem

If the population of all the magazine subscribers were normal, you would expect its sampling distribution of means to be normal as well. But what if the population were non-normal? The **central limit theorem** states that even if a population distribution is strongly non-normal, its sampling distribution of means will be approximately normal for large sample sizes (over 30). The central limit theorem makes it possible to use probabilities associated with the normal curve to answer questions about the means of sufficiently large samples.

Note that the larger the sample, the less variable the sample mean. The mean of many observations is less variable than the mean of few. The standard deviation of a sampling distribution of means is often called the **standard error** of the mean. According to the central limit theorem, the mean of a sampling distribution of means is an unbiased estimator of the population mean.

$$\mu_{\bar{x}} = \mu$$

Similarly, the standard deviation of a sampling distribution of means is

$$\sigma_{\bar{x}} = \frac{\sigma}{\sqrt{n}}$$

Every statistic has a standard error, which is a measure of the statistic's random variability.

Example 1: If the population mean of number of fish caught per trip to a particular fishing hole is 3.2 and the population standard deviation is 1.8, what are the mean and standard deviation of the distribution with a sample size of 40 trips?

$$\mu_{\bar{x}} = 3.2$$

$$\sigma_{\bar{x}} = \frac{1.8}{\sqrt{40}} = 0.285$$

Properties of the Normal Curve

Known characteristics of the normal curve make it possible to estimate the probability of occurrence of any value of a normally distributed variable. To determine a normal distribution, you need to know its mean and standard deviation. Suppose that the total area under the curve is defined to be 1. You can multiply that number by 100 and say there is a 100 percent chance that any value you can name will be somewhere in the distribution. (***Remember:*** The distribution extends to infinity in both directions.) Similarly, because half the area of the curve is below the mean and half is above it, you can say there is a 50 percent chance that a randomly chosen value will be above the mean and the same chance that it will be below it.

It makes sense that the area under the normal curve is equivalent to the probability of randomly drawing a value in that range. The area is greatest in the middle, where the "hump" is, and thins out toward the tails. That is consistent with the fact that there are more values close to the mean in a normal distribution than far from it.

When the area of the standard normal curve is divided into sections by standard deviations above and below the mean, the area in each section is a known quantity (see the following figure). As explained earlier, the area in each section corresponds to the probability of randomly drawing a value in that range.

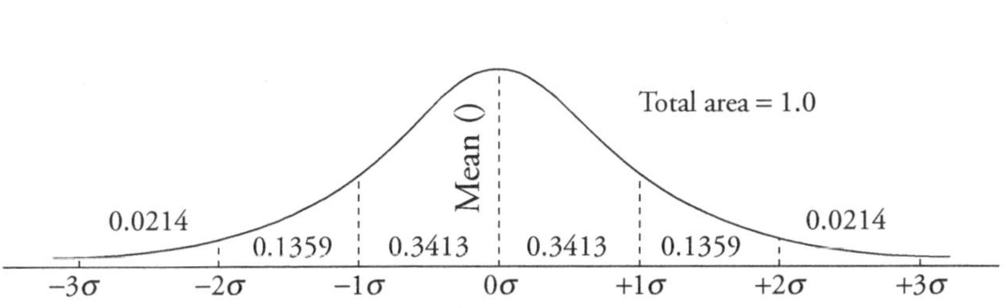

For example, 0.3413 of the curve falls between the mean and one standard deviation above the mean, which means that about 34 percent of all the values of a normally distributed variable are between the mean and one standard deviation above it. It also means that there is a 0.3413 chance that a value drawn at random from the distribution will lie between these two points.

Sections of the curve above and below the mean may be added together to find the probability of obtaining a value within (plus or minus) a given number of standard deviations of the mean (see the following figure). For example, the amount of curve area between one standard deviation above the mean and one standard deviation below—that is, within one standard deviation—is 0.3413 + 0.3413 = 0.6826, which means that approximately 68.26 percent of the values lie in that range. Similarly, about 95 percent of the values lie within two standard deviations of the mean, and 99.7 percent of the values lie within three standard deviations.

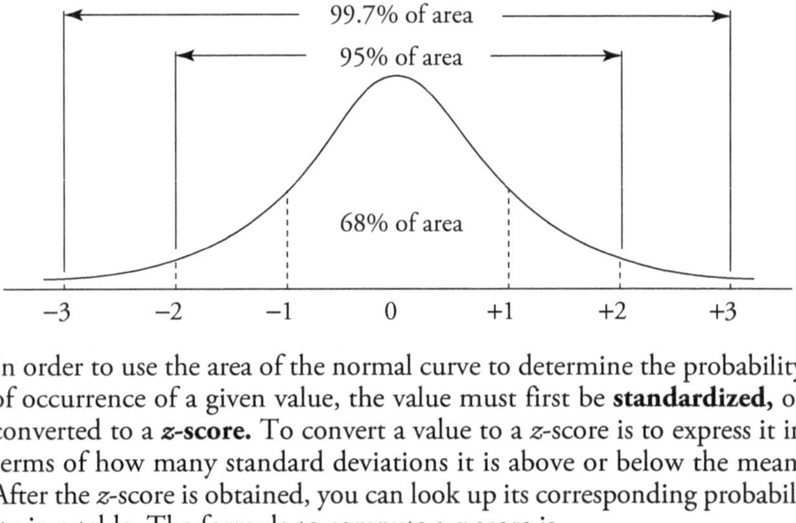

In order to use the area of the normal curve to determine the probability of occurrence of a given value, the value must first be **standardized,** or converted to a ***z*-score.** To convert a value to a *z*-score is to express it in terms of how many standard deviations it is above or below the mean. After the *z*-score is obtained, you can look up its corresponding probability in a table. The formula to compute a *z*-score is

$$z = \frac{x - \mu}{\sigma}$$

where x is the value to be converted, μ is the population mean, and σ is the population standard deviation.

Example 2: A normal distribution of retail-store purchases has a mean of \$14.31 and a standard deviation of 6.40. What percentage of purchases were under \$10? First, compute the *z*-score:

$$z = \frac{10 - 14.31}{6.40} = -\ 0.67$$

The next step is to look up the *z*-score in the table of standard normal probabilities (see Table B-2 in Appendix B). The standard normal table lists the probabilities (curve areas) associated with given *z*-scores.

Table B-2 gives the area of the curve below *z*—in other words, the probability of obtaining a value of *z* or lower. Not all standard normal tables use the same format, however. Some list only positive *z*-scores and give the area of the curve between the mean and *z*. Such a table is slightly more difficult to use, but the fact that the normal curve is symmetric makes it possible to use it to determine the probability associated with any *z*-score, and vice versa.

To use Table B-2 (the table of standard normal probabilities), first look up the *z*-score in the left column, which lists *z* to the first decimal place. Then look along the top row for the second decimal place. The intersection of the row and column is the probability. For Example 2, you first find −0.6 in the left column and then 0.07 in the top row. Their intersection is 0.2514. The answer, then, is that about 25 percent of the purchases were under \$10 (see the following figure).

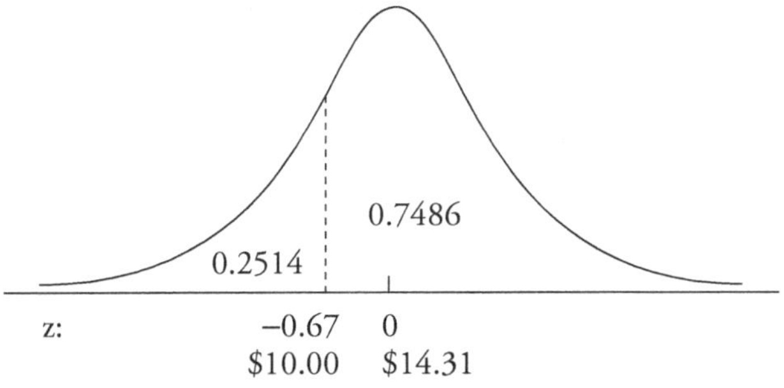

What if you wanted to know the percentage of purchases above a certain amount? Because Table B-2 gives the area of the curve below a given *z*, to obtain the area of the curve above *z*, simply subtract the tabled probability from 1. The area of the curve above a *z* of −0.67 is 1 − 0.2514 = 0.7486. Approximately 75 percent of the purchases were above \$10.

Just as Table B-2 can be used to obtain probabilities from *z*-scores, it can be used to do the reverse, as in Example 3.

Example 3: Using the previous example, what purchase amount marks the lower 10 percent of the distribution?

In Table B-2, locate the probability of 0.1000, or as close as you can find, and read off the corresponding *z*-score. The figure that you seek lies between the tabled probabilities of 0.0985 and 0.1003, but closer to 0.1003, which corresponds to a *z*-score of -1.28. Now, use the *z* formula, this time solving for *x*:

$$\frac{x - 14.31}{6.4} = -1.28$$

$$x - 14.31 = (-1.28)(6.4)$$

$$x = -8.192 + 14.31$$

$$= 6.118$$

Approximately 10 percent of the purchases were below $6.12.

Normal Approximation to the Binomial

In Chapter 4, "Probability," you saw that some variables are continuous—there is no limit to the number of times you could divide their intervals into still smaller ones, although you may round them off for convenience. Examples include age, height, and cholesterol level. Other variables are discrete, or made of whole units with no values between them. Some discrete variables are the number of children in a family, the sizes of televisions available for purchase, or the number of medals awarded at the Olympic Games.

Chapter 4 also introduced one kind of discrete variable: the binomial variable. A binomial variable can take only two values, often termed *successes* and *failures.* Examples include coin tosses that come up either heads or tails, manufactured parts that either continue working past a certain point or do not, and basketball tosses that either fall through the hoop or do not.

You discovered that the outcomes of binomial trials have a frequency distribution, just as continuous variables do. The more binomial trials there are (for example, the more coins you toss simultaneously), the more closely the sampling distribution resembles a normal curve (see the figure on p. 71). You can take advantage of this fact and use the table of standard normal probabilities (Table B-2) to estimate the likelihood of obtaining a given proportion of successes. You can do this by converting the test proportion to a *z*-score and looking up its probability in the standard normal table.

As shown in the following figure, as the number of trials increases, the binomial distribution approaches the normal distribution.

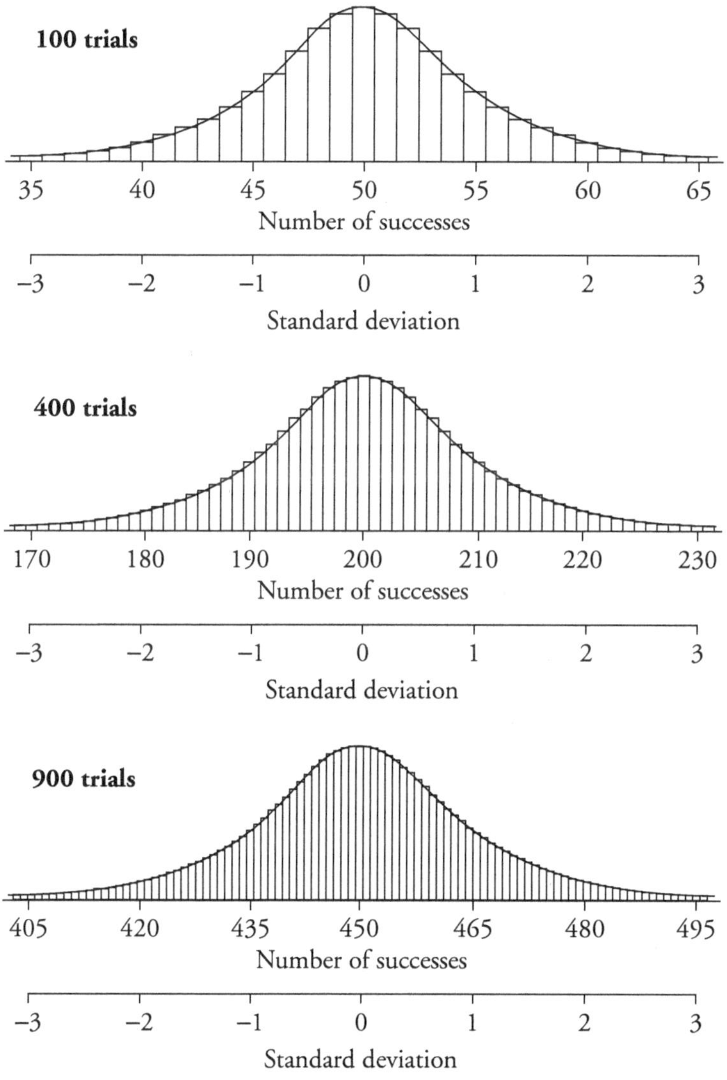

The mean of the normal approximation to the binomial is

$$\mu = n\pi$$

and the standard deviation is

$$\sigma = \sqrt{n\pi(1-\pi)}$$

where n is the number of trials and π is the probability of success. The approximation will be more accurate the larger the n and the closer the proportion of successes in the population is to 0.5.

Example 4: Assuming an equal chance of a new baby being a boy or a girl (that is, $\pi = 0.5$), what is the likelihood that more than 60 out of the next 100 births at a local hospital will be boys?

$$\begin{aligned} z &= \frac{x-\mu}{\sigma} \\ &= \frac{60-((100)(0.5))}{\sqrt{(100)(0.5)(1-0.5)}} \\ &= \frac{10}{5} \\ &= 2 \end{aligned}$$

According to Table B-2, a z-score of 2 corresponds to a probability of 0.9772. As you can see in the following figure, there is a 0.9772 chance that there will be 60 percent or fewer boys, which means that the probability that there will be more than 60 percent boys is $1 - 0.9772 = 0.0228$, or just over 2 percent. If the assumption that the chance of a new baby being a girl is the same as it being a boy is correct, the probability of obtaining 60 or fewer girls in the next 100 births is also 0.9772.

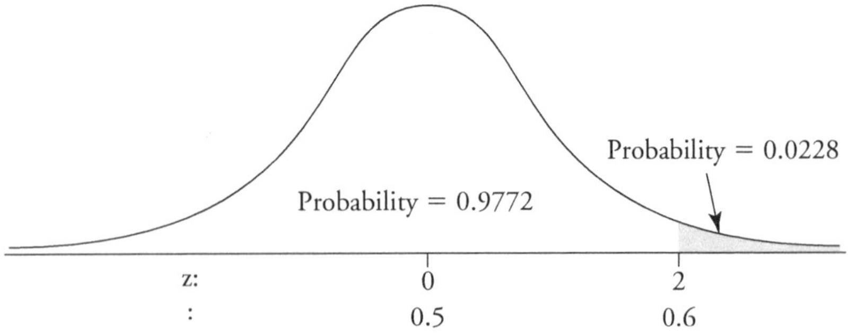

Chapter Check-Out

Questions

1. A population has a mean of 500 and a standard deviation of 100. What is the probability that a randomly selected sample from this population, of size 25, will have a mean greater than 530?
2. The speed limit on the street in front of your house is 40 mph. In order to measure compliance with the limit, you decide to measure the mean speed of vehicles on the street. Each day after work, you measure the speed of 20 cars and take their average. Your data for 5 days is shown in the following table.

Day	***Average Speed (mph)***
Monday	38.3
Tuesday	40.1
Wednesday	34.2
Thursday	45.1
Friday	50.7

   Using this information, answer the following questions:

   (a) Find the mean and standard deviation of these means. Is it likely that drivers comply, on average, with the law?

   (b) Assuming the population mean of drivers does comply, what is the probability of your results? Take $\sigma = 7.5$ mph. Assume a normal distribution of speeds about μ.

3. On TV, surveys are often taken during sensational news stories to determine the "people's opinion." Usually, viewers are told to dial a phone number in order to vote yes or no on the issue. These phone numbers sometimes require the caller to pay a fee. Comment on this in terms of sampling practice.

   (a) Does it produce a random sample?

   (b) If not, how could it be modified to produce a more random sample?

Answers

1. 0.0668
2. **(a)** mean: 41.68, standard deviation of the means: 6.38, yes; **(b)** 0.59
   Hint: The word "comply" should let you know that the mean is <40. Use the normal distribution and related formulas in this chapter to find the probability.
3. **(a)** no; **(b)** random people should be called by the station

Chapter 6

PRINCIPLES OF HYPOTHESIS TESTING

Chapter Check-In

- ❑ Stating hypotheses and corresponding null hypotheses
- ❑ Learning how to apply simple statistical tests to determine if a null hypothesis is rejected
- ❑ Understanding the types of statistical errors
- ❑ Calculating statistical significances, point estimates, and confidence intervals
- ❑ Estimating a difference score

Common Core Standard: Making Inferences and Justifying Conclusions

Make inferences and justify conclusions from sample surveys, experiments, and observational studies. Use data from a randomized experiment to compare two treatments; use simulations to decide if differences between parameters are significant. Evaluate reports based on data (S.IC).

Most people assume that the whole point of statistical analysis is to prove or disprove a certain claim. It is not usually possible to prove or disprove the claim, but it is possible to *establish evidence* to test a claim. This chapter introduces you to techniques for testing claims based on the available data. These techniques are simple yet surprisingly powerful. Common Core Mathematics suggests that you learn how to analyze your decision outcomes and provide evidence for the claimed results to see if they are consistent with the results from a given data-generated process. Making decisions and providing evidence requires you to learn about hypothesis testing.

Stating a Hypothesis

One common use of statistics is the testing of scientific hypotheses. A **hypothesis** is a reasonable guess that is made about one of a population's parameters; for example, the mean or standard deviation. To set the stage for the hypothesis testing process, it is necessary to have a research hypothesis, a null hypothesis, and a significance level.

First, the investigator forms a **research hypothesis** that states an expectation to be tested. It is what the investigator predicts will happen during the research study. Then the investigator derives a statement that is the opposite of the research hypothesis. This statement is called the **null hypothesis** (in notation, H_0), which is actually the default assumption. It is the null hypothesis that is actually tested, not the research hypothesis. If the null hypothesis can be rejected, that is taken as evidence in favor of the research hypothesis (also called the **alternative hypothesis** (in notation, H_a). Because individual tests are rarely conclusive, it is usually not said that the research hypothesis has been "proven," only that it has been supported. Rather, as a conclusion, investigators will either reject the null hypothesis in favor of the research hypothesis or not reject the null hypothesis.

An example of a research hypothesis comparing two groups might be the following:

> Fourth graders at Elmwood School perform differently in math from fourth graders at Lancaster School. This could be measured by comparing the means of these groups.
>
> In notation, H_a: $\mu_1 \neq \mu_2$ or sometimes H_a: $\mu_1 - \mu_2 \neq 0$.

The null hypothesis would be:

> Fourth graders at Elmwood School perform the same in math as fourth graders at Lancaster School.
>
> In notation, H_0: $\mu_1 = \mu_2$ or H_0: $\mu_1 - \mu_2 = 0$.

Some research hypotheses are more specific than that, predicting not only a difference, but also a difference in a particular direction. These are often described as one-sided tests:

> Fourth graders at Elmwood School are *better* in math than fourth graders at Lancaster School.
>
> In notation, H_a: $\mu_1 > \mu_2$ or H_a: $\mu_1 - \mu_2 > 0$.

The Test Statistic

Hypothesis testing involves the use of distributions of known area, like the normal distribution, to estimate the probability of obtaining a certain value as a result of chance. The researcher is usually testing to see if the probability will be low because that means it is likely that the test result was not a mere coincidence but occurred because the researcher's theory is correct. It could mean, for example, that it is probably not just bad luck but faulty packaging equipment that caused you to get a box of raisin cereal with only five raisins in it.

Only two outcomes of a hypothesis test are possible: Either the null hypothesis is rejected, or it is not. You have seen that values from normally distributed populations can be converted to *z*-scores and their probabilities looked up in Table B-2 in Appendix B (see Chapter 5 for a complete discussion). The *z*-score is one kind of **test statistic** that is used to determine the probability of obtaining a given value. In order to test hypotheses, you must decide in advance what number to use as a cutoff for whether or not the null hypothesis will be rejected. This number is called the **critical value** (sometimes called the **tabled value** because it is looked up in a table). It represents the level of probability that you will use to test the hypothesis. If the computed test statistic has a smaller probability than that of the critical value, the null hypothesis will be rejected.

For example, suppose you want to test the theory that sunlight helps prevent depression. One hypothesis derived from this theory might be that hospital admission rates for depression in sunny regions of the country are lower than the national average. Suppose that you know the national annual admission rate for depression to be 17 per 10,000. You intend to take the mean of a sample of admission rates from hospitals in sunny parts of the country and compare it to the national average.

Your research hypothesis is:

> The mean annual admission rate for depression from the hospitals in sunny areas is less than 17 per 10,000.
>
> In notation, H_a: $\mu_1 < 17$ per 10,000.

The null hypothesis is:

> The mean annual admission rate for depression from the hospitals in sunny areas is equal to 17 per 10,000.
>
> In notation, H_0: $\mu_1 = 17$ per 10,000.

Your next step is to choose a probability level for the test. You know that the sample mean must be lower than 17 per 10,000 in order to reject the null hypothesis, but how much lower? You settle on a probability level of 5 percent. That is, if the mean admission rate for the sample of hospitals in sunny areas is so low that the chance of obtaining that rate from a sample selected at random from the national population is less than 5 percent, you will reject the null hypothesis and conclude there is evidence to support the hypothesis that exposure to the sun reduces the incidence of depression.

Next, you look up the critical *z*-score—the *z*-score that corresponds to your chosen level of probability—in the standard normal table. It is important to remember which end of the distribution you are concerned with. Table B-2 lists the probability of obtaining a given *z*-score or lower. That is, it gives the area of the curve below the *z*-score. Because a computed test statistic in the lower end of the distribution will allow you to reject your null hypothesis, you look up the *z*-score for the probability (or area) of 0.05 and find that it is −1.65. If you were hypothesizing that the mean in sunny parts of the country is greater than the national average, you would have been concerned with the upper end of the distribution instead and would have looked up the *z*-score associated with the probability (area) of 0.95, which is 1.65.

The critical *z*-score allows you to define the **region of acceptance** and the **region of rejection** of the curve (see the figure that follows). If the computed test statistic is below the critical *z*-score, you can reject the null hypothesis and say that you have provided evidence in support of the alternative hypothesis. If it is above the critical z-score, you cannot reject the null hypothesis.

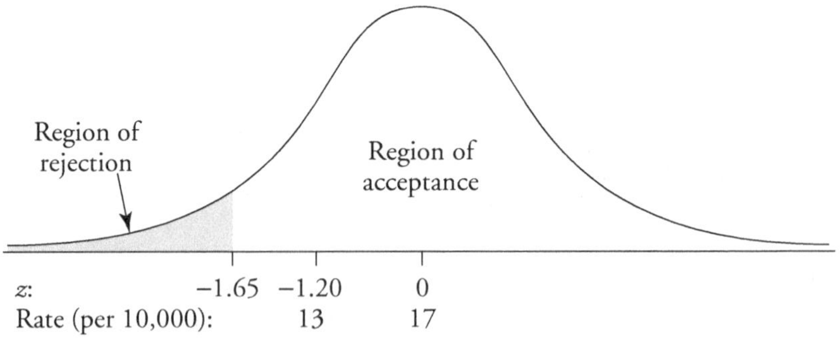

Suppose that the mean admission rate for the sample of hospitals in sunny regions is 13 per 10,000 and suppose also that the corresponding *z*-score for that mean is −1.20. The test statistic falls in the region of acceptance, so you cannot reject the null hypothesis that the mean in sunny parts of the country is significantly lower than the mean in the national average. There is a greater than 5 percent chance of obtaining a mean admission rate of 13 per 10,000 or lower from a sample of hospitals chosen at random from the national population, so you cannot conclude that your sample mean could not have come from that population.

One- and Two-Tailed Tests

In the previous example, you tested a research hypothesis that predicted not only that the sample mean would be different from the population mean, but also that it would be different in a specific direction—it would be lower. This test is called a **directional** or **one-tailed test** because the region of rejection is entirely within one tail of the distribution.

Some hypothesis predict only that one value will be different from another, without additionally predicting which will be higher. The test of such a hypothesis is **nondirectional** or **two-tailed** because an extreme test statistic in either tail of the distribution (positive or negative) will lead to the rejection of the null hypothesis of no difference.

Suppose that you suspect that a particular class's performance on a proficiency test is not representative of those people who have taken the test. The national mean score on the test is 74.

The research hypothesis is:

> The mean score of the class on the test is not 74.
>
> In notation, H_a: $\mu \neq 74$.

The null hypothesis is:

> The mean score of the class on the test is 74.
>
> In notation, H_0: $\mu = 74$.

As in the hospital example, you decide to use a 5 percent probability level for the test. Both tests have a region of rejection, then, of 5 percent, or 0.05. In this example, however, the rejection region must be split between both tails of the distribution—0.025 in the upper tail and 0.025 in the lower tail—because your hypothesis specifies only a difference, not a direction, as shown in graph (a) below. You will reject the null hypothesis of no difference if the class sample mean is either much higher or much lower than the population mean of 74. In the hospital example, however, only a sample mean much lower than the population mean would have led to the rejection of the null hypothesis.

Two Tests at the Same Probability Level (95%)

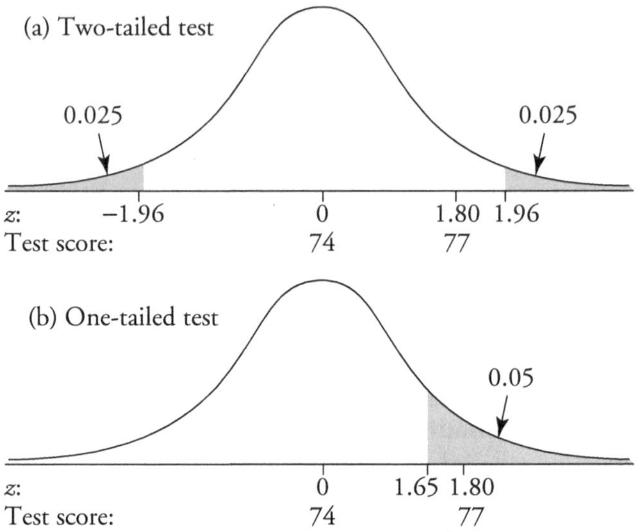

The decision of whether to use a one- or a two-tailed test is important because a test statistic that falls in the region of rejection in a one-tailed test may not do so in a two-tailed test, even though both tests use the same probability level. Suppose the class sample mean in your example was 77 and its corresponding *z*-score was computed to be 1.80. Table B-2 in Appendix B shows the critical *z*-scores for a probability of 0.025 in either tail to be −1.96 and 1.96. In order to reject the null hypothesis, the test statistic must be either smaller than −1.96 or greater than 1.96. It is not, so you cannot reject the null hypothesis. Refer to graph (a) above.

Suppose, however, you had a reason to expect that the class would perform better on the proficiency test than the population, and you did a one-tailed test instead. For this test, the rejection region of 0.05 would be entirely within the upper tail. The critical *z*-value for a probability of 0.05 in the upper tail is 1.65. (Remember that Table B-2 gives areas of the curve below *z*, so you look up the *z*-value for a probability of 0.95.) Your computed test statistic of $z = 1.80$ exceeds the critical value and falls in the region of rejection, so you reject the null hypothesis and say that your suspicion that the class was better than the population was supported. See graph (b) in the figure above.

In practice, you should use a one-tailed test only when you have good reason to expect that the difference will be in a particular direction. A two-tailed test is more conservative than a one-tailed test because a two-tailed test requires a more extreme test statistic to reject the null hypothesis.

Type I and II Errors

You have been using probability to decide whether a statistical test provides evidence for or against your predictions. If the likelihood of obtaining a given test statistic from the population is very small, you reject the null hypothesis and say you have supported your hunch that the sample you are testing is different from the population.

But you could be wrong. Even if you choose a probability level of 5 percent, that means there is a 5 percent chance, or 1 in 20, that you rejected the null hypothesis when it was, in fact, correct. You can err in the opposite way, too; you might fail to reject the null hypothesis when it is, in fact, incorrect. These two errors are called Type I and Type II, respectively. Table 6-1 presents the four possible outcomes of any hypothesis test based on (1) whether the null hypothesis was accepted or rejected, and (2) whether the null hypothesis was true.

Table 6-1 Types of Statistical Errors

	H_0 is actually:	
	True	***False***
Reject H_0	Type I error	Correct
Accept H_0	Correct	Type II error

A **Type I error** is often represented by the Greek letter alpha (α) and a **Type II error** by the Greek letter beta (β). In choosing a level of probability for a test, you are actually deciding how much you want to risk committing a Type I error—rejecting the null hypothesis when it is, in fact, true. For this reason, the area in the region of rejection is sometimes called the *alpha level* because it represents the likelihood of committing a Type I error.

In order to graphically depict a Type II, or β, error, it is necessary to imagine next to the distribution for the null hypothesis a second distribution for the true alternative (see the following figure). If the alternative hypothesis is actually true, but you fail to reject the null hypothesis for all values of the test statistic falling to the left of the critical value, then the area of the curve of the alternative (true) hypothesis lying to the left of the critical value represents the percentage of times that you will have made a Type II error.

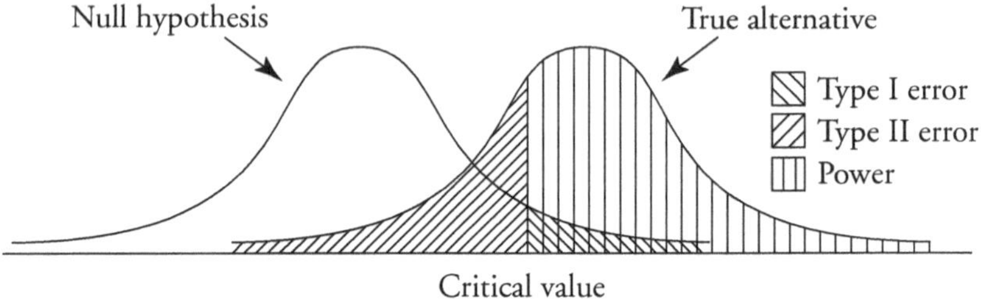

Type I and Type II errors are inversely related: As the rate of one increases, the rate of the other decreases. The Type I, or α (alpha), error rate is usually set in advance by the researcher. The Type II error rate for a given test is harder to know because it requires estimating the distribution of the alternative hypothesis, which is usually unknown.

A related concept is **power**—the probability that a test will reject the null hypothesis when it is, in fact, false. You can see from the figure above that power is simply 1 minus the Type II error rate (β). High power is desirable. Like β, power can be difficult to estimate accurately, but increasing the sample size always increases power.

Statistical Significance

How do you know how much confidence to put in the outcome of a hypothesis test? The statistician's criterion is the **statistical significance** of the test, or the likelihood of obtaining a given result by chance. The result is said to be statistically significant if it is sufficient to reject the null hypothesis. This concept has been spoken of already using several terms: probability, area of the curve, Type I error rate, and so forth. Another common representation of significance is the letter *p* (for probability) and a number between 0 and 1. There are several ways to refer to the significance level of a test, and it is important to be familiar with them. All of the following statements, for example, are equivalent:

The finding is significant at the 0.05 level.

The **confidence level** is 95 percent.

The Type I error rate is 0.05.
The alpha level is 0.05.
$\alpha = 0.05$.
There is a 1 in 20 chance of obtaining this result.
The area of the region of rejection is 0.05.
The p-value is 0.05.
$p = 0.05$.

The smaller the significance level p, the more stringent the test and the greater the likelihood that the conclusion is correct. The significance level is usually chosen in consideration of other factors that affect and are affected by it, like sample size, estimated size of the effect being tested, and consequences of making a mistake. Common significance levels are 0.10 (1 chance in 10), 0.05 (1 chance in 20), and 0.01 (1 chance in 100).

The result of a hypothesis test, as has been seen, is that the null hypothesis is either rejected or not rejected. The significance level for the test is set in advance by the researcher in choosing a critical test value. When the computed test statistic is large (or small) enough for the tester to reject the null hypothesis, however, it is customary to report the observed (actual) p-value for the statistic.

If, for example, you intend to perform a one-tailed (lower tail) test using the standard normal distribution at $p = 0.05$, the test statistic will have to be smaller than the critical z-value of -1.65 in order to reject the null hypothesis. But suppose the computed z-score is -2.50, which has an associated probability of 0.0062. The null hypothesis is rejected with room to spare. The observed significance level of the computed statistic is $p = 0.0062$, so you could report that the result was significant at $p < 0.01$. This result means that even if you had chosen the more stringent significance level of 0.01 in advance, you still would have rejected the null hypothesis, which is stronger support for your research hypothesis than rejecting the null hypothesis at $p = 0.05$.

It is important to realize that statistical significance and substantive, or practical, significance are not the same thing. A small but important real-world difference may fail to reach significance in a statistical test. Conversely, a statistically significant finding may have no practical consequence. This finding is especially important to remember when working with large sample sizes because any difference can be statistically significant if the samples are extremely large.

Point Estimates and Confidence Intervals

A **confidence interval** (CI) is an interval that gives us a good estimated range of values of unknown population parameters. In other words, it is the degree of confidence we have that the estimated interval, constructed through many separate data analysis processes (after repeating the experiment many times), contains the true values of the population parameter that is being tested. The most frequently used percentages for confidence intervals are 90 percent, 95 percent, and 99 percent. For example, a 90 percent confidence interval means that "we are 90 percent confident that the true value of the parameter is in our confidence interval." The confidence interval is the complement of the respective level of significance. For example, a 90 percent confidence interval means a 0.10 significance level.

You have seen that the sample mean $\bar{x}$ is an unbiased estimate of the population mean μ. Another way to say this is that $\bar{x}$ is the best **point estimate** of the true value of μ. Some error is associated with this estimate, however—the true population mean may be larger or smaller than the sample mean. Instead of a point estimate, you might want to identify a range of possible values p might take, controlling the probability that μ is not lower than the lowest value in this range and not higher than the highest value. Such a range is called a *confidence interval.*

Example 1: Suppose that you want to find out the average weight of all players on the football team at Landers College. You are able to select ten players at random and weigh them. The mean weight of the sample of players is 198 pounds, so that number is your point estimate. Assume that the population standard deviation is $\sigma = 11.50$. What is a 90 percent confidence interval for the population weight if you presume the players' weights are normally distributed?

This question is the same as asking what weight values correspond to the upper and lower limits of an area of 90 percent in the center of the distribution. You can define that area by looking up the *z*-scores in Table B-2 in Appendix B that correspond to probabilities of 0.05 in either end of the distribution. They are -1.65 and 1.65. You can determine the weights that correspond to these *z*-scores using the following formula:

$$(a,b) = \bar{x} \pm z \cdot \frac{\sigma}{\sqrt{n}}$$

The weight values for the lower and upper ends of the confidence interval are 192 and 204 (see the figure on p. 85). A confidence interval is usually expressed by two values enclosed by parentheses, as in (192, 204). Another way to express the confidence interval is as the point estimate plus or minus a **margin of error**; in this case, it is 198 ± 6 pounds. You are 90 percent certain that the true population mean of football player weights is between 192 and 204 pounds.

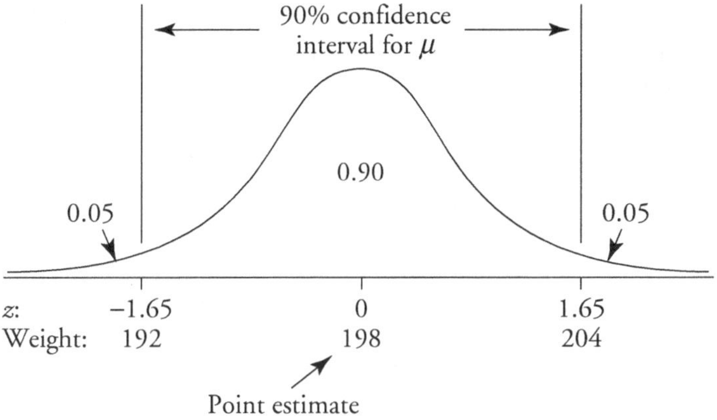

What would happen to the confidence interval if you wanted to be 95 percent certain of it? You would have to draw the limits (ends) of the intervals closer to the tails in order to encompass an area of 0.95 between them instead of 0.90. That would make the low value lower and the high value higher, which would make the interval wider. The width of the confidence interval is related to the confidence level, standard error, and *n* such that the following are true:

- The higher the percentage of confidence desired, the wider the confidence interval.
- The larger the standard error, the wider the confidence interval.
- The larger the *n*, the smaller the standard error, and so the narrower the confidence interval.

All other things being equal, a smaller confidence interval is always more desirable than a larger one because a smaller interval means the population parameter can be estimated more accurately.

Estimating a Difference Score

Imagine that instead of estimating a single population mean μ, you wanted to estimate the difference between two population means μ_1 and μ_2, such as the difference between the mean weights of two football teams. The statistic

$\bar{x}_1 - \bar{x}_2$ has a sampling distribution just as the individual means do, and the rules of statistical inference can be used to calculate either a point estimate or a confidence interval for the difference between the two population means.

Suppose you wanted to know which was greater, the mean weight of Landers College's football team or the mean weight of Ingram College's football team. You already have a point estimate of 198 pounds for the Landers team. Suppose that you draw a random sample of players from the Ingram team, and the sample mean is 195 pounds. The point estimate for the difference between the mean weights of the Landers team (μ_1) and the Ingram team (μ_2) is 198 − 195 = 3.

But how accurate is that estimate? You can use the sampling distribution of the difference score to construct a confidence interval for $\mu_1 - \mu_2$, just as you did before. Suppose that when you do so, you find that the confidence interval limits are (−3, 9), which means you are 90 percent certain that the mean for the Landers team is between 3 pounds lighter and 9 pounds heavier than the mean for the Ingram team (see the following figure).

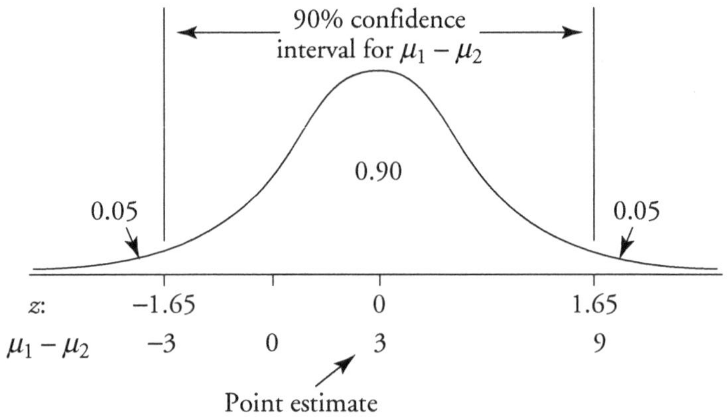

Suppose instead of a confidence interval, you want to test the two-tailed hypothesis that the two team weights have different means. Your null hypothesis would be:

$$H_0: \mu_1 = \mu_2 \text{ or } H_0: \mu_1 - \mu_2 = 0$$

To reject the null hypothesis of equal means, the test statistic—in this example, the *z*-score—for a difference in mean weights of 0 would have to fall in the rejection region at either end of the distribution. But you have already seen that it does not—only difference scores less than −3 or greater than 9 fall in the rejection region. For this reason, you would be unable to reject the null hypothesis that the two population means are equal.

This characteristic is a simple but important one of confidence intervals for difference scores. If the interval contains 0, you would be unable to reject the null hypothesis that the means are equal at the same significance level.

Univariate Tests: An Overview

To summarize, hypothesis testing of problems with one variable requires carrying out the following steps:

1. State the null hypothesis and the alternative hypothesis.
2. Decide on a significance level for the test.
3. Compute the value of a test statistic.
4. Compare the test statistic to a critical value from the appropriate probability distribution corresponding to your chosen level of significance and observe whether the test statistic falls within the region of acceptance or the region of rejection. Equivalently, compute the *p*-value that corresponds to the test statistic and compare it to the selected significance level.

Thus far, you have used the test statistic z and the table of standard normal probabilities (Table B-2 in Appendix B) to carry out your tests. There are other test statistics and other probability distributions. The general formula for computing a test statistic for making an inference about a single population is

$$\text{test statistic} = \frac{\text{observed sample statistic} - \text{hypothesized value}}{\text{standard error}}$$

where *observed sample statistic* is the statistic of interest from the sample (usually the mean), *hypothesized value* is the hypothesized population parameter (again, usually the mean), and *standard error* is the standard deviation of the sampling distribution divided by the positive square root of n.

The general formula for computing a test statistic for making an inference about a difference between two populations is

$$\text{test statistic} = \frac{\text{statistic}_1 - \text{statistic}_2 - \text{hypothesized value}}{\text{standard error}}$$

where $statistic_1$ and $statistic_2$ are the statistics from the two samples (usually the means) to be compared, *hypothesized value* is the hypothesized difference between the two population parameters (0 if testing for equal values), and *standard error* is the standard error of the sampling distribution, whose formula varies according to the type of problem.

The general formula for computing a confidence interval is

$$\text{observed sample statistic} \pm \text{critical value} \times \text{standard error}$$

where *observed sample statistic* is the point estimate (usually the sample mean), *critical value* is from the table of the appropriate probability distribution (upper or positive *z*-value) corresponding to half the desired alpha level, and *standard error* is the standard error of the sampling distribution.

Why must the alpha level be halved before looking up the critical value when computing a confidence interval? Because the rejection region is split between both tails of the distribution, as in a two-tailed test. For a confidence interval at $\alpha = 0.05$, you would look up the critical value corresponding to an upper-tailed probability of 0.025.

Chapter Check-Out

Questions

1. An advertiser wants to know if the average age of people regularly watching a particular TV show is less than 24 years old.
   (a) Is this a one- or two-tailed test?
   (b) A random survey of 50 viewers determines that their mean age is 19 years, with a standard deviation of 1.7 years. Find the 90 percent confidence interval of the age of the viewers.
2. True or False: Statistical tests should always be performed on a null hypothesis.
3. True or False: A result with a high level of significance is always very important.
4. A UCLA statistics professor claims that his coin is fair. If you want to test his claim, what would be the null hypothesis?
5. In the following claim, is the test one-tailed or two-tailed?
   "The average score of the final exam in an anthropology class is less than 62."

Answers

1. **(a)** one-tailed; **(b)** (18.6, 19.4)
2. True
3. False
4. H_0: P = 0.5
5. one-tailed

Chapter 7

UNIVARIATE INFERENTIAL TESTS

Chapter Check-In

- ❑ Learning about one-sample statistical tests, including the *z*-test and *t*-test
- ❑ Learning about comparing two means in two-sample *z*-tests and *t*-tests
- ❑ Calculating confidence levels and confidence intervals for each test
- ❑ Learning about tests for a single population proportion and comparing two proportions
- ❑ Applying these tests to examples

Common Core Standard: Making Inferences and Justifying Conclusions

Decide if a specified model is consistent with results from a given data-generating process. Evaluate reports based on data (S.IC). Understand that a randomized experiment can be used to compare two treatments. Use statistics appropriate to the shape of the data distribution to compare two or more data sets (S.ID.2).

Chapter 6 explained how to formulate hypotheses and test them with the *z*-test. In Chapter 6, you also saw how a general test statistic is constructed and can be used to test a hypothesis against any probability distribution.

Common Core Mathematics requires you to learn about different statistical tests to be able to compare results of two given data tests or two different groups in a set of observations. In this chapter, you take a closer look at two of the most common statistical tests used in scientific and sociological research: the *z*-test and the *t*-test. These tests are used to test hypotheses concerning population means or proportions. Each test is explained in a separate section with a simple format so that you can easily look up how to calculate the appropriate test statistic. The formula for finding the confidence interval of your data is given as well. Most useful of all, each section contains several simple examples to illustrate how to apply the test to your data.

> ***Note:*** In this chapter, when rounding rules are not stated in the problem, round to the nearest one-hundredth.

One-Sample *z*-test

The ***z*-test** is used to estimate the mean of a population and compare it with the reference value when the standard deviation of the population is given. It helps us to understand whether the mean is different from a specific value, or to calculate a range of the values that may include the population mean. ***Note:*** The *z*-test works very well with sample sizes that are 30 or larger because the central limit theorem justifies the sample standard deviation as a suitable approximation to the population standard deviation.

Requirements: Normally distributed population, σ known

Test for population mean: Hypothesis test

Formula: $z = \dfrac{\overline{x} - \mu}{\dfrac{\sigma}{\sqrt{n}}}$

where $\overline{x}$ is the sample mean, μ is the population mean, σ is the population standard deviation, and n is the size of the sample. Look up the significance level of the *z*-value in the standard normal (z) table (Table B-2 in Appendix B).

Example 1 (one-tailed test): A herd of 1,500 steer was fed a special high-protein grain for a month. A random sample of 29 were weighed and had gained an average of 6.7 pounds. If the standard deviation of weight gain for the entire herd is 7.1 pounds, test the hypothesis that the average weight gain per steer for the month was more than 5 pounds.

Null hypothesis: H_0: $\mu = 5$

Alternative hypothesis: H_a: $\mu > 5$

$$z = \frac{6.7 - 5}{\frac{7.1}{\sqrt{29}}} = \frac{1.7}{1.318437} = 1.289$$

Rounded to the nearest one-hundredth, the tabled value for $z \leq 1.29$ is 0.9015.

$$1 - 0.9015 = 0.0985$$

So, the conditional probability that a sample from the herd gains at least 6.7 pounds per steer is $p = 0.0985$. Should the null hypothesis of a weight gain of less than 5 pounds for the population be rejected? That depends on how conservative you want to be. If you had decided beforehand on a significance level of $p < 0.05$, the null hypothesis could not be rejected.

Example 2 (two-tailed test): In national use, a vocabulary test is known to have a mean score of 68 and a standard deviation of 13. A class of 19 students takes the test and has a mean score of 65. Is the class' performance typical of that of others who have taken the test? Assume a significance level of $p < 0.05$.

There are two possible ways that the class may differ from the population: Its scores may be lower than or higher than the population of all students taking the test. Therefore, this problem requires a two-tailed test. First, state the null and alternative hypotheses:

Null hypothesis: H_0: $\mu = 68$

Alternative hypothesis: H_a: $\mu \neq 68$

Because you have specified a significance level, you can look up the critical z-value in Table B-2 before computing the statistic. This is a two-tailed test, so the 0.05 must be split such that 0.025 is in the upper tail and another 0.025 in the lower tail. The z-value that corresponds to -0.025 is -1.96, which is the lower critical z-value. The upper value corresponds to $1 - 0.025$, or 0.975, which gives a z-value of 1.96. The null hypothesis of no difference will be rejected if the computed z statistic falls outside the range of -1.96 to 1.96.

Next, compute the z statistic:

$$z = \frac{65 - 68}{\frac{13}{\sqrt{19}}} = \frac{-3}{2.982} = -1.006$$

Because -1.006 is between -1.96 and 1.96, the null hypothesis of population mean is 68 and cannot be rejected. That is, there is no evidence that this class' performance can be considered different from that of others who have taken the test.

Confidence interval for population mean using *z*

Formula: $(a, b) = \bar{x} \pm z_{a/2} \cdot \frac{\sigma}{\sqrt{n}}$

where a and b are the limits of the confidence interval, $\bar{x}$ is the sample mean, $z_{\alpha/2}$ is the upper (or positive) *z*-value from the standard normal (*z*) table (Table B-2) corresponding to half of the desired alpha level (because all confidence intervals are two-tailed), σ is the population standard deviation, and n is the size of the sample.

Example 3: A sample of 12 machine pins has a mean diameter of 1.15 inches, and the population standard deviation is known to be 0.04. What is a 99 percent confidence interval of diameter width for the population?

First, determine the *z*-value. A 99 percent confidence level is equivalent to $p < 0.01$. Half of 0.01 is 0.005. The *z*-value corresponding to an area of 0.005 is 2.58. The interval may now be calculated:

$$1.15 \pm 2.58 \cdot \frac{0.04}{\sqrt{12}} = 1.15 \pm 0.03$$

The interval is (1.12, 1.18).

We have 99 percent confidence that the population mean of pin diameters lies between 1.12 and 1.18 inches. Note that this is not the same as saying that 99 percent of the machine pins have diameters between 1.12 and 1.18 inches, which would be an incorrect conclusion from this test.

Choosing a sample size

Because surveys cost money to administer, researchers often want to calculate how many subjects will be needed to determine a population mean using a fixed confidence interval and significance level. The formula is

$$n = \left(\frac{z_{a/2}\sigma}{w} \right)^2$$

where n is the number of subjects needed, $z_{\alpha/2}$ is the critical z-value corresponding to the desired significance level, σ is the population standard deviation, and w is the desired confidence interval width.

Example 4: How many subjects will be needed to find the average age of students at Fisher College, plus or minus 1 year, with a 95 percent significance level and a population standard deviation of 3.5?

$$n = \left(\frac{(1.96)(3.5)}{2}\right)^2 = \left(\frac{6.86}{2}\right)^2 = 11.76$$

Rounding up, a sample of 12 students would be sufficient to determine students' mean age plus or minus 1 year. Note that the confidence interval width is always double the "plus or minus" figure.

One-Sample *t*-test

As discussed in the previous section, the *z*-test is used when both the mean and the standard deviation of a null hypothesis population (σ) are known. However, it is not always practical to use the *z*-test because the value of the population is not usually known. In place of the *z*-test, the *t*-test is widely used in behavioral sciences to estimate the sample standard deviation (s). The ***t*-test** is a distribution for estimating the mean of a normally distributed data set when sample size is small and the population standard deviation is unknown. The *t*-test is also used for finding the confidence intervals and the statistical significance of the difference of two means. The graph of a *t*-test distribution looks like a normal distribution but has heavier tails, meaning *there are more chances to produce values that are far from the mean.* Note that the higher the degree of freedom, the closer the *t*-distribution is to the normal curve.

Requirements: Normally distributed population; σ is unknown

Test for population mean: Hypothesis test

Formula: $t = \dfrac{\overline{x} - \mu}{\dfrac{s}{\sqrt{n}}}$

where $\overline{x}$ is the sample mean, μ is the population mean (a specified value to be tested), s is the sample standard deviation, and n is the size of the sample. Notice that the *t*-test uses the standard deviation of the sample (s), whereas the z-test uses the standard deviation of the null hypothesis population (σ).

When the standard deviation of the sample is substituted for the standard deviation of the population, the statistic does not have a normal distribution; it has what is called the ***t*-distribution** (see Table B-3 in Appendix B). Because there is a different t-distribution for each sample size, it is not practical to list a separate area-of-the-curve table for each one. Instead, critical t-values for common alpha levels (0.10, 0.05, 0.01, and so forth) are usually given in a single table for a range of sample sizes. For very large samples, the t-distribution approximates the standard normal (z) distribution. In practice, it is best to use t-distributions any time the population standard deviation is not known.

Values in the t-table are not actually listed by sample size but by **degrees of freedom** (df). The number of degrees of freedom for a problem involving the t-distribution for sample size n is simply $n - 1$ for a one-sample mean problem.

Example 5 (one-tailed test): A professor wants to know if the students in her introductory statistics class have a good grasp of basic math. Six students are chosen at random from the class and given a math proficiency test. The professor wants the class to be able to score above 70 on the test. The six students get scores of 62, 92, 75, 68, 83, and 95. Can the professor have 90 percent confidence that the mean score for the class on the test would be above 70?

Null hypothesis: H_0: $\mu = 70$

Alternative hypothesis: H_a: $\mu > 70$

First, compute the sample mean and standard deviation (see Chapter 3):

$$\begin{array}{r} 62 \\ 92 \\ 75 \\ 68 \\ 83 \\ +\ 95 \\ \hline 475 \end{array} \qquad \begin{aligned} \overline{x} &= \frac{475}{6} = 79.17 \\ s &= 13.17 \end{aligned}$$

Next, compute the *t*-value:

$$t = \frac{79.17 - 70}{\frac{13.17}{\sqrt{6}}} = \frac{9.17}{5.38} = 1.70$$

To test the hypothesis, the computed *t*-value of 1.70 will be compared to the critical value in the *t*-table (Table B-3). But which do you expect to be larger and which do you expect to be smaller? One way to approach this is to look at the formula and see what effect different means would have on the computation. If the sample mean had been 85 instead of 79.17, the resulting *t*-value would have been larger. Because the sample mean is in the numerator, the larger it is, the larger the resulting figure will be. At the same time, you know that a higher sample mean will make it more likely the professor will conclude that the math proficiency of the class is satisfactory and that the null hypothesis of less-than-satisfactory class math knowledge can be rejected. Therefore, it must be true that the larger the computed *t*-value, the greater the chance that the null hypothesis can be rejected. It follows, then, that if the computed *t*-value is larger than the critical *t*-value from the table, the null hypothesis can be rejected.

A 90 percent confidence level is equivalent to an alpha level of 0.10. Because extreme values in one rather than two directions will lead to rejection of the null hypothesis, this is a one-tailed test, and you do not divide the alpha level by 2. The number of degrees of freedom for the problem is 6 − 1 = 5. The value in the *t*-table (Table B-3) for $t_{.10,\,5}$ is 1.476. Because the computed *t*-value of 1.70 is larger than the critical value in the table, the null hypothesis can be rejected, and the professor has evidence that the class mean on the math test would be at least 70.

Note that the formula for the one-sample *t*-test for a population mean is the same as the *z*-test, except that the *t*-test substitutes the sample standard deviation s for the population standard deviation σ and takes critical values from the *t*-distribution instead of the *z*-distribution. The *t*-distribution is particularly useful for tests with small samples ($n < 30$).

Example 6 (two-tailed test): A Little League baseball coach wants to know if his team is representative of other teams in scoring runs. Nationally, the average number of runs scored by a Little League team in a game is 5.7. He chooses five games at random in which his team scored 5, 9, 4, 11, and 8 runs. Is it likely that his team's scores could have come from the national distribution? Assume an alpha level of 0.05.

Because the team's scoring rate could be either higher than or lower than the national average, the problem calls for a two-tailed test. First, state the null and alternative hypotheses:

Null hypothesis: H_0: $\mu = 5.7$

Alternative hypothesis: H_a: $\mu \neq 5.7$

Next, compute the sample mean and standard deviation:

$$\begin{array}{r} 5 \\ 9 \\ 4 \\ 11 \\ \underline{+8} \\ 37 \end{array} \qquad \bar{x} = \frac{37}{5} = 7.4 \qquad s = 2.88$$

Next, compute the *t*-value:

$$t = \frac{7.4 - 5.7}{\frac{2.88}{\sqrt{5}}} = \frac{1.7}{1.29} = 1.32$$

Now, look up the critical value from the *t*-table (Table B-3). You need to know two things in order to do this: the degrees of freedom and the desired alpha level. The degrees of freedom is 5 − 1 = 4. The overall alpha level is 0.05, but because this is a two-tailed test, the alpha level must be divided by two, which yields 0.025. The tabled value for $t_{.025,\,4}$ is 2.776. The computed *t* of 1.32 is smaller than the *t* from Table B-3, so you cannot reject the null hypothesis that the mean of this team is equal to the population mean. The coach cannot conclude that his team is different from the national distribution on runs scored.

Confidence interval for population mean using *t*

Formula: $(a,\ b) = \bar{x} \pm t_{a/2,\,df} \cdot \frac{s}{\sqrt{n}}$

where *a* and *b* are the limits of the confidence interval, $\bar{x}$ is the sample mean, $t_{\alpha/2,\,df}$ is the value from the *t*-table (Table B-3) corresponding to half of the desired alpha level at $n - 1$ degrees of freedom, *s* is the sample standard deviation, and *n* is the size of the sample.

Example 7: Using the information in Example 6, what is a 95 percent confidence interval for runs scored per team per game?

First, determine the *t*-value. A 95 percent confidence level is equivalent to an alpha level of 0.05. Half of 0.05 is 0.025. The *t*-value corresponding to an area of 0.025 at either end of the *t*-distribution for 4 degrees of freedom ($t_{.025,\,4}$) is 2.776. The interval may now be calculated:

$$\begin{aligned}(a,\ b) &= 7.4 \pm 2.78\left(\frac{2.88}{\sqrt{5}}\right)\\ &= 7.4 \pm 3.59\\ &= (3.81,\ 10.99)\end{aligned}$$

The interval is fairly wide, mostly because *n* is small.

Two-Sample *z*-test for Comparing Two Means

Requirements: Two normally distributed but independent populations; σ is known.

Test for population mean: Hypothesis test

Formula: $z = \dfrac{\overline{x}_1 - \overline{x}_2 - \mu}{\sqrt{\dfrac{\sigma_1^2}{n_1} + \dfrac{\sigma_2^2}{n_2}}}$

where $\overline{x}_1$ and $\overline{x}_2$ are the means of the two samples, μ is the hypothesized difference between the population means (0 if testing for equal means), σ_1 and σ_2 are the standard deviations of the two populations, and n_1 and n_2 are the sizes of the two samples.

Example 8 (two-tailed test): The amount of a certain trace element in blood is known to vary with a standard deviation of 14.1 ppm (parts per million) for male blood donors and 9.5 ppm for female blood donors. Random samples of 75 male donors and 50 female donors yield concentration means of 28 and 33 ppm, respectively. What is the likelihood that the population means of concentrations of the element are the same for men and women?

Null hypothesis: H_0: $\mu_1 = \mu_2$ or H_0: $\mu_1 - \mu_2 = 0$

Alternative hypothesis: $H_a\colon \mu_1 \neq \mu_2$ or $H_a\colon \mu_1 - \mu_2 \neq 0$

$$z = \frac{28 - 33 - 0}{\sqrt{\frac{14.1^2}{75} + \frac{9.5^2}{50}}} = \frac{-5}{\sqrt{2.65 + 1.81}} = -2.37$$

The computed z-value is negative because the (larger) mean for females was subtracted from the (smaller) mean for males. But because the hypothesized difference between the populations is 0, the order of the samples in this computation is arbitrary—$\bar{x}_1$ could just as well have been the female sample mean and $\bar{x}_2$ the male sample mean, in which case z would be 2.37 instead of −2.37. An extreme z-score in either tail of the distribution (plus or minus) will lead to rejection of the null hypothesis of no difference.

The area of the standard normal curve corresponding to a z-score of −2.37 is 0.0089. Because this test is two-tailed, that figure is doubled to yield a probability of 0.0178 that the population means are the same. If the test had been conducted at a prespecified significance level of $\alpha < 0.05$, the null hypothesis of equal means could be rejected. If the specified significance level had been the more conservative (more stringent) $\alpha < 0.01$, however, the null hypothesis could not be rejected.

In practice, the two-sample z-test is not used often because the two population standard deviations, σ_1 and σ_2, are usually unknown. Instead, sample standard deviations and the t-distribution are used.

Two-Sample *t*-test for Comparing Two Means

Requirements: Two normally distributed but independent populations; σ is unknown.

Test for population mean: Hypothesis test

Formula: $$t = \frac{\bar{x}_1 - \bar{x}_2 - \mu}{\sqrt{\frac{s_1^2}{n_1} + \frac{s_2^2}{n_2}}}$$

where $\bar{x}_1$ and $\bar{x}_2$ are the means of the two samples, μ is the hypothesized difference between the population means (0 if testing for equal means),

s_1 and s_2 are the standard deviations of the two samples, and n_1 and n_2 are the sizes of the two samples. The number of degrees of freedom for the problem is the smaller of $n_1 - 1$ and $n_2 - 1$.

Example 9 (one-tailed test): An experiment is conducted to determine whether intensive tutoring (covering a great deal of material in a fixed amount of time) is more effective than paced tutoring (covering less material in the same amount of time). Two randomly chosen groups are tutored separately and then administered proficiency tests. Use a significance level of $\alpha < 0.05$.

Let μ_1 represent the population mean for the intensive tutoring group and μ_2 represent the population mean for the paced tutoring group.

Null hypothesis: H_0: $\mu_1 = \mu_2$ or H_0: $\mu_1 - \mu_2 = 0$

Alternative hypothesis: H_a: $\mu_1 > \mu_2$ or H_a: $\mu_1 - \mu_2 > 0$

Group	Method	n	$\bar{x}$	s
1	Intensive	12	46.31	6.44
2	Paced	10	42.79	7.52

$$t = \frac{46.31 - 42.79 - 0}{\sqrt{\frac{6.44^2}{12} + \frac{7.52^2}{10}}} = \frac{3.52}{\sqrt{3.46 + 5.66}} = 1.166 = 1.17$$

The degrees of freedom parameter is the smaller of (12 − 1) and (10 − 1), or 9. Because this is a one-tailed test, the alpha level (0.05) is not divided by two. The next step is to look up $t_{.05,\,9}$ in the t-table (Table B-3 in Appendix B), which gives a critical value of 1.833. The computed t of 1.17 does not exceed the tabled value, so the null hypothesis cannot be rejected. This test has not provided statistically significant evidence that intensive tutoring is superior to paced tutoring.

Confidence interval for comparing two means

Formula: $(a,\ b) = \bar{x}_1 - \bar{x}_2 \pm t_{a/2,\,df} \cdot \sqrt{\frac{s_1^2}{n_1} + \frac{s_2^2}{n_2}}$

where a and b are the limits of the confidence interval, $\bar{x}_1$ and $\bar{x}_2$ are the means of the two samples, $t_{\alpha/2,\,df}$ is the value from the t-table (Table B-3) corresponding to half of the desired alpha level, s_1 and s_2 are the standard deviations of the two samples, and n_1 and n_2 are the sizes of the two samples. The degrees of freedom parameter for looking up the t-value is the smaller of $n_1 - 1$ and $n_2 - 1$.

Example 10: Estimate a 90 percent confidence interval for the difference between the number of raisins per box in two brands of breakfast cereal.

Brand	**n**	**$\bar{x}$**	**s**
A	6	102.1	12.3
B	9	93.6	7.52

The difference between $\bar{x}_1$ and $\bar{x}_2$ is 102.1 − 93.6 = 8.5. The degrees of freedom is the smaller of (6 − 1) and (9 − 1), or 5. A 90 percent confidence interval is equivalent to an alpha level of 0.10, which is then halved to give 0.05. According to Table B-3, the critical value for $t_{.05,\,5}$ is 2.015. The interval may now be computed.

$$8.5 \pm 2.015 \cdot \sqrt{\frac{12.3^2}{6} + \frac{7.52^2}{9}} = 8.5 \pm 2.015 \cdot \sqrt{25.22 + 6.28}$$

$$= 8.5 \pm 11.31 \text{ (rounded to the nearest 100th)}$$

The interval is (−2.81, 19.81).

You can be 90 percent confident that Brand A cereal has between 2.81 fewer and 19.81 more raisins per box than Brand B. The fact that the interval contains 0 means if you had performed a test of the hypothesis that the two population means are different (using the same significance level), you would not have been able to reject the null hypothesis of no difference.

Pooled variance method

If the two population distributions can be assumed to have the same variance—and, therefore, the same standard deviation—s_1 and s_2 can be pooled together, each weighted by the number of cases in their respective samples. Although using pooled variance in a t-test is generally more likely to yield significant results than using separate variances, it is often hard to know whether the variances of the two populations are equal. For this reason, the pooled variance method should be used with caution. The formula for the pooled estimator of σ^2 is

$$s_p^2 = \frac{(n_1 - 1)s_1^2 + (n_2 - 1)s_2^2}{n_1 + n_2 - 2}$$

where s_1 and s_2 are the standard deviations of the two samples and n_1 and n_2 are the sizes of the two samples.

The formula for comparing the means of two populations using pooled variance is

$$t = \frac{\bar{x}_1 - \bar{x}_2 - \mu}{\sqrt{s_p^2\left(\frac{1}{n_1} + \frac{1}{n_2}\right)}}$$

where $\bar{x}_1$ and $\bar{x}_2$ are the means of the two samples, μ is the hypothesized difference between the population means (0 if testing for equal means), s_p^2 is the pooled variance, and n_1 and n_2 are the sizes of the two samples. The number of degrees of freedom for the problem is $df = n_1 + n_2 - 2$.

Example 11 (two-tailed test): Does right- or left-handedness affect how fast people type? Random samples of students from a typing class are given a typing speed test (words per minute), and the results are compared. Significance level for the test is 0.10. Because you are looking for a difference between the groups in either direction (right-handed faster than left-handed, or vice versa), this is a two-tailed test.

Null hypothesis: H_0: $\mu_1 = \mu_2$ or H_0: $\mu_1 - \mu_2 = 0$

Alternative hypothesis: H_a: $\mu_1 \neq \mu_2$ or H_a: $\mu_1 - \mu_2 \neq 0$

Group	***-Handed***	***n***	$\bar{x}$	***s***
1	Right	16	55.8	5.7
2	Left	9	59.3	4.3

First, calculate the pooled variance:

$$\begin{aligned} s_p^2 &= \frac{(16-1)5.7^2 + (9-1)4.3^2}{16+9-2} \\ &= \frac{487.35 + 147.92}{23} \\ &= 27.62 \end{aligned}$$

Next, calculate the t-value:

$$t = \frac{55.8 - 59.3 - 0}{\sqrt{27.62\left(\frac{1}{16} + \frac{1}{9}\right)}} = \frac{-3.5}{\sqrt{4.70}} = \frac{-3.5}{2.17} = -1.61$$

The degrees of freedom parameter is 16 + 9 − 2, or 23. This test is a two-tailed one, so you divide the alpha level (0.10) by two. Next, you look up $t_{.05,\,23}$ in the t-table (Table B-3), which gives a critical value of 1.714. This value is larger than the absolute value of the computed t of −1.61, so the null hypothesis of equal population means cannot be rejected. There is no evidence that right- or left-handedness has any effect on typing speed.

Paired Difference *t*-test

Requirements: A set of paired observations from a normal population

This t-test compares one set of measurements with a second set from the same sample. It is often used to compare "before" and "after" scores in experiments to determine whether significant change has occurred; for example, comparing the weights of a group of students before and after exercising three times a week.

Another application of the t-test might be when we want to compare two different groups from a gathered set of data; for example, when comparing the preferred method of teaching a statistics topic. Another example might be if you are teaching adults how to use Facebook and Twitter, and then later comparing the two applications to observe which social network was preferred and more user friendly.

Test for population mean: Hypothesis test

Formula: $t = \dfrac{\overline{x} - \mu}{\dfrac{s}{\sqrt{n}}}$

where $\overline{x}$ is the mean of the change scores, μ is the hypothesized difference (0 if testing for equal means), s is the sample standard deviation of the differences, and n is the sample size. The number of degrees of freedom for the problem is $n - 1$.

Example 12 (one-tailed test): A farmer decides to try out a new fertilizer on a test plot containing 10 stalks of corn. Before applying the fertilizer, he measures the height of each stalk. Two weeks later, he measures the stalks again, being careful to match each stalk's new height to its previous one. The stalks would have grown an average of 6 inches during that time even without the fertilizer. Did the fertilizer help? Use a significance level of 0.05.

Null hypothesis: H_0: $\mu = 6$

Alternative hypothesis: H_a: $\mu > 6$

Stalk	*1*	*2*	*3*	*4*	*5*	*6*	*7*	*8*	*9*	*10*
Before height	35.5	31.7	31.2	36.3	22.8	28.0	24.6	26.1	34.5	27.7
After height	45.3	36.0	38.6	44.7	31.4	33.5	28.8	35.8	42.9	35.0

Subtract each stalk's "before" height from its "after" height to get the change score for each stalk; then, compute the mean and standard deviation of the change scores and insert these into the formula.

$$45.3 - 35.5 = 9.8$$
$$36.0 - 31.7 = 4.3$$
$$38.6 - 31.2 = 7.4$$
$$44.7 - 36.3 = 8.4$$
$$31.4 - 22.8 = 8.6$$
$$33.5 - 28.0 = 5.5$$
$$28.8 - 24.6 = 4.2$$
$$35.8 - 26.1 = 9.7$$
$$42.9 - 34.5 = 8.4$$
$$35.0 - 27.7 = 7.3$$
$$\frac{73.6}{10} = 7.36$$

$$s^2 = \frac{(9.8-7.36)^2 + (4.3-7.36)^2 + \cdots + (7.3-7.36)^2}{10-1}$$
$$= 4.22$$
$$s = \sqrt{4.22}$$
$$= 2.05$$
$$t = \frac{7.36-6}{\frac{2.05}{\sqrt{10}}} = \frac{1.36}{0.65} = 2.09$$

The problem has $n - 1$, or $10 - 1 = 9$ degrees of freedom. The test is one-tailed because you are asking only whether the fertilizer increases growth, not reduces it. The critical value from the *t*-table (Table B-3 in Appendix B) for $t_{.05,\,9}$ is 1.833.

Because the computed *t*-value of 2.09 is larger than 1.833, the null hypothesis can be rejected. The test has provided evidence that the fertilizer caused the corn to grow more than if it had not been fertilized. The amount of actual increase was not large (1.36 inches over normal growth), but it was statistically significant.

Test for a Single Population Proportion

Requirements: Binomial population, sample $n\pi_0 \geq 10$, and sample $n(1 - \pi_0) \geq 10$, where π_0 is the hypothesized proportion of successes in the population

Test for population mean: Hypothesis test

Formula: $$z = \frac{\hat{\pi} - \pi_0}{\sqrt{\dfrac{\pi_0\left(1 - \pi_0\right)}{n}}}$$

where $\hat{\pi}$ is the sample proportion, π_0 is the hypothesized proportion, and n is the sample size. Because the distribution of sample proportions is approximately normal for large samples, the z statistic is used. The test is most accurate when π (the population proportion) is close to 0.5 and least accurate when π is close to 0 or 1.

Example 13 (one-tailed test): The sponsors of a city marathon have been trying to encourage more women to participate in the event. A sample of 70 runners is taken, of which 32 are women. The sponsors would like to be 90 percent certain that at least 40 percent of the participants are women. Were their recruitment efforts successful?

Null hypothesis: H_0: $\pi = 0.4$

Alternative hypothesis: H_a: $\pi > 0.4$

The proportion of women runners in the sample is 32 out of 70, or 45.7 percent. The *z*-value may now be calculated:

$$z = \frac{0.457 - 0.40}{\sqrt{\dfrac{0.40(1 - 0.40)}{70}}} = \frac{0.057}{\sqrt{0.00343}} = 0.97$$

From the standard normal (z) table, you find that the probability of a z-value less than 0.97 is 0.834. Therefore, the p-value is $1 - 0.834 = 0.166$, so we do not reject the null hypothesis. Thus, it cannot be concluded at that level of significance that the population of runners is at least 40 percent women.

Confidence interval for a single population proportion

Formula: $(a,\ b) = \hat{\pi} \pm z_{a/2} \cdot \sqrt{\frac{\hat{\pi}(1-\hat{\pi})}{n}}$

where $\hat{\pi}$ is the sample proportion, $z_{\alpha/2}$ is the upper z-value corresponding to half of the desired alpha level, and n is the sample size.

Example 14: A sample of 100 voters selected at random in a congressional district prefer Candidate Smith to Candidate Jones by a ratio of 3 to 2. What is a 95 percent confidence interval of the percentage of voters in the district who prefer Smith?

A ratio of 3 to 2 is equivalent to a proportion of $\frac{3}{5} = 0.60$. A 95 percent confidence interval is equivalent to an alpha level of 0.05, half of which is 0.025. The critical z-value corresponding to an upper probability of $1 - 0.025$ is 1.96. The interval may now be computed:

$$\begin{aligned}(a,\ b) &= 0.60 \pm 1.96\sqrt{\frac{0.60(1-0.60)}{100}} \\ &= 0.60 \pm 1.96\sqrt{0.002} \\ &= 0.60 \pm 0.096 \\ &= (0.504,\ 0.696)\end{aligned}$$

We have 95 percent confidence that between 50.4 percent and 69.6 percent of the voters in the district prefer Candidate Smith. Note that the problem could have been figured for Candidate Jones by substituting the proportion 0.40 for Smith's proportion of 0.60.

Choosing a sample size

In Example 14, you estimated that the percentage of voters in the district who prefer Candidate Smith is 60 percent plus or minus about 10 percent. Another way to say this is that the estimate has a standard **margin of error** of ± 10 percent, or a confidence interval width of 20 percent. That is a pretty wide range. You may wish to make the margin smaller.

Because the width of the confidence interval decreases at a known rate as the sample size increases, it is possible to determine the sample size needed to estimate a proportion with a fixed confidence interval. The formula is

$$n = \left(\frac{2z_{\alpha/2}}{w}\right)^2 \cdot p^*\left(1-p^*\right)$$

where n is the number of subjects needed, $z_{\alpha/2}$ is the z-value corresponding to half of the desired significance level, w is the desired confidence interval width, and p^* is an estimate of the true population proportion. A p^* of 0.50 will result in a higher n than any other proportion estimate but is often used when the true proportion is not known.

Example 15: How large a sample is needed to estimate the preference of district voters for Candidate Smith with a margin of error of ± 4 percent at a 95 percent significance level?

You will conservatively estimate the (unknown) true population proportion of preference for Smith at 0.50. If it is really larger (or smaller) than that, you will overestimate the size of the sample needed, but $p^* = 0.50$ is playing it safe.

$$\begin{aligned} n &= \left(\frac{2 \cdot 1.96}{0.08}\right)^2 \cdot 0.50(1-0.50) \\ &= 49^2 \cdot 0.25 \\ &= 600.25 \end{aligned}$$

A sample of about 601 voters would be needed both to estimate the percentage of voters in the district who prefer Smith, and in order to be 95 percent certain that the estimate is within ±4 percent of the true population percentage.

Test for Comparing Two Proportions

Requirements: Two binomial populations; $n\pi_0 \geq 5$ and $n\,(1 - \pi_0) \geq 5$ (for the respective samples), where π_0 is the hypothesized proportion of successes in the population

Difference test: Hypothesis test

Formula: $$z = \frac{\hat{\pi}_1 - \hat{\pi}_2 - \mu}{\hat{\pi}\left(1-\hat{\pi}\right)\left(\frac{1}{n_1} + \frac{1}{n_2}\right)}$$

where $\hat{\pi} = \dfrac{x_1 + x_2}{n_1 + n_2}$ and where $\hat{\pi}_1$ and $\hat{\pi}_2$ are the sample proportions, μ is their hypothesized difference (0 if testing for equal proportions), n_1 and n_2 are the sample sizes, and x_1 and x_2 are the number of "successes" in each sample. As in the test for a single proportion, the *z*-distribution is used to test the hypothesis.

Example 16 (one-tailed test): A swimming school wants to determine whether a recently hired instructor is working out. Sixteen out of 25 of Instructor A's students passed the lifeguard certification test on the first try. In comparison, 57 out of 72 of the more experienced Instructor B's students passed the test on the first try. Is Instructor A's success rate worse than Instructor B's? Use $\alpha = 0.10$.

Null hypothesis: H_0: $\pi_1 = \pi_2$

Alternative hypothesis: H_a: $\pi_1 < \pi_2$

First, you need to compute the values for some of the terms in the formula.

The sample proportion $\hat{\pi}_1$ is $\dfrac{16}{25} = 0.640$. The sample proportion $\hat{\pi}_2$ is $\dfrac{57}{72} = 0.792$. Next, compute $\hat{\pi}$:

$$\hat{\pi} = \frac{16+57}{25+72} = \frac{73}{97} = 0.753$$

Finally, compute the main formula:

$$\begin{aligned} z &= \frac{0.640 - 0.792 - 0}{\sqrt{0.753(1-0.753)\left(\frac{1}{25}+\frac{1}{72}\right)}} \\ &= \frac{-0.152}{\sqrt{(0.186)(0.054)}} \\ &= \frac{-0.152}{0.100} \\ &= -1.52 \end{aligned}$$

The standard normal (*z*) table (Table B-2 in Appendix B) shows that the lower critical *z*-value for $\alpha = 0.10$ is approximately -1.28. The computed

z must be lower than -1.28 in order to reject the null hypothesis of equal proportions. Because the computed z is -1.52, the null hypothesis can be rejected. It can be concluded (at this level of significance) that Instructor A's success rate is worse than Instructor B's.

Confidence interval for comparing two proportions

Formula: $(a,\ b)=\hat{\pi}_1-\hat{\pi}_2\pm z_{a/2}\cdot s(D)$

where $s(D)=\sqrt{\dfrac{\hat{\pi}_1(1-\pi_1)}{n_1}+\dfrac{\hat{\pi}_2(1-\pi_2)}{n_2}}$ and where a and b are the limits of the confidence interval of $\pi_1 - \pi_2$, $\hat{\pi}_1$ and $\hat{\pi}_2$ are the sample proportions, $z_{\alpha/2}$ is the upper z-value corresponding to half of the desired alpha level, and n_1 and n_2 are the sizes of the two samples.

Example 17: A public health researcher wants to know how two high schools—one in the inner city and one in the suburbs—differ in the percentage of students who smoke. A random survey of students gives the following results:

Population	***n***	***Smokers***
1 (inner city)	125	47
2 (suburban)	153	52

What is a 90 percent confidence interval for the difference between the smoking rates in the two schools?

The proportion of smokers in the inner-city school is $\hat{\pi}_1=\dfrac{47}{125}=0.376$.

The proportion of smokers in the suburban school is $\hat{\pi}_2=\dfrac{52}{153}=0.340$. Next, solve for $s(D)$:

$$\begin{aligned}s(D)&=\sqrt{\frac{0.376(1-0.376)}{125}+\frac{0.340(1-0.340)}{153}}\\&=\sqrt{\frac{0.235}{125}+\frac{0.224}{153}}\\&=\sqrt{0.00334}\\&=0.06\text{ (rounded to the nearest 100th)}\end{aligned}$$

A 90 percent confidence interval is equivalent to $\alpha = 0.10$, which is halved to give 0.05. The upper tabled value for $z_{.05}$ is 1.65. The interval may now be computed:

$$\begin{aligned}(a,\ b) &= (0.376 - 0.340) \pm (1.65)(0.06) \\ &= 0.036 \pm 0.099 \\ &= (-0.06,\ 0.14)\end{aligned}$$

The researcher can be 90 percent confident that the true population proportion of smokers in the inner-city high school is between 6 percent lower and 14 percent higher than the proportion of smokers in the suburban high school. Thus, since the confidence interval contains 0, there is no significant difference between the two types of schools at $\alpha = 0.10$.

Chapter Check-Out

Questions

1. Suppose you count the number of days in May on which it rains in your town and find it to be 15 out of 31. Checking the weather database, you find that for your town, the average number of rainy days in May is 10.

(a) What type of test would you use to determine if this May was unusually wet to $\alpha = 0.05$?

(b) Perform the test. What is the value of your test statistic (that is, z or t)?

2. A biologist believes that polluted water may be causing frogs in a lake to be smaller than normal. In an unpolluted lake, she measures 15 adult frogs and finds a mean length of 7.6 inches with a standard deviation of 1.5 inches. In the polluted lake, she measures 23 adult frogs and finds a mean of 6.5 inches and a standard deviation of 2.3 inches.

(a) State the null and alternative hypotheses.

(b) Calculate the appropriate test statistic.

3. A high school track coach makes the sprinters on his team begin weight lifting in the hopes that their speed will increase. Before beginning, each runs a 40-yard dash. After a month of weight training, each runs the dash again. Their times are given below.

Runner	*1*	*2*	*3*	*4*	*5*	*6*	*7*	*8*	*9*	*10*
Before training	4.8	5.1	5.5	4.9	5.6	6.0	5.8	5.3	6.1	4.7
After training	4.7	5.0	5.1	5.0	5.2	5.1	5.6	5.1	5.5	5.0

(a) State the null and alternative hypotheses.
(b) To what level can the coach be sure that his runners have improved?

4. Select one of the following tests for the real-world scenarios described in questions 4(a), 4(b), and 4(c).

I. paired-difference t-test
II. two-sample t-test
III. one-sample t-test

(a) Which is the preferred test to compare the average time that is needed for students in classes A and B to answer a statistics final exam?
(b) A group of students are given an IQ test before and after taking a geometry class. Which is the preferred test to determine if taking a geometry class helps to increase or decrease the IQ test results?
(c) Which is the preferred test to compare the mean of the age of the voters in Arkansas who voted for Candidate X with the mean of the voters who voted for Candidate X in the entire country?

Answers

1. **(a)** test for single population proportion; **(b)** 1.92
2. **(a)** H_0: $\mu_{\text{pol}} \geq \mu_{\text{unpol}}$; H_a: $\mu_{\text{pol}} < \mu_{\text{unpol}}$; **(b)** 1.784
3. **(a)** H_0: $\mu_{\text{after}} \geq \mu_{\text{before}}$; H_a: $\mu_{\text{after}} < \mu_{\text{before}}$; **(b)** $p < 0.025$
4. **(a)** II. two-sample t-test; **(b)** I. paired-difference t-test; **(c)** III. one-sample t-test

Chapter 8
BIVARIATE RELATIONSHIPS

Chapter Check-In

- ❑ Checking for correlation between two variables
- ❑ Learning about regression
- ❑ Using the chi-square (χ^2) test to determine whether two variables are independent

Common Core Standard: Interpreting Categorical and Quantitative Data

Summarize, represent, and interpret data (S.ID). Interpret the slope (rate of change) and the intercept (constant term) of a linear model in the context of data (S.ID.7). Distinguish between correlation and causation (S.ID.9). Make inferences and justify conclusions (S.IC).

So far, you have been working with problems involving a single variable. Many times, however, you may want to know something about the relationship between two variables. Common Core Mathematics suggests that you learn how to study the possible relationships between variables, find the correlation (if there is any), interpret the slope (rate of change), find the intercept (constant term) of a linear model in the context of the data, compute and interpret the correlation coefficient of a linear fit, and distinguish between correlation and causation. This chapter explains all of these **bivariate** (two-variable) relationships.

Correlation

Consider Table 8-1, which contains measurements of two variables for ten people: the number of months the person has owned an exercise machine and the number of hours the person spent exercising in the past week.

Table 8-1 Exercise Data for Ten People

Person	*1*	*2*	*3*	*4*	*5*	*6*	*7*	*8*	*9*	*10*
Months Owned	5	10	4	8	2	7	9	6	1	12
Hours Exercised	5	2	8	3	8	5	5	7	10	3

If you display these data pairs as points in a scatter plot (see figure below), you can see a definite trend. The points appear to form a line that slants from the upper left to the lower right. As you move along that line from left to right, the values on the vertical axis (hours of exercise) get smaller, while the values on the horizontal axis (months owned) get larger. Another way to express this is to say that the two variables are inversely related: The more months the machine was owned, the less the person tended to exercise.

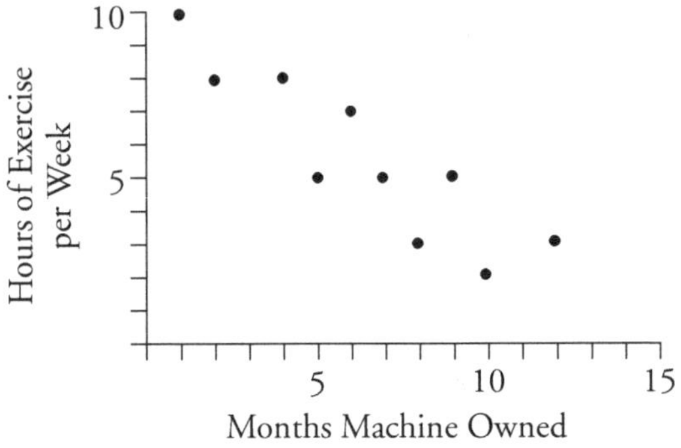

These two variables have a **correlation.** More than that, they are correlated in a particular direction—negatively. For an example of a positive correlation, suppose that instead of displaying "hours of exercise" on the vertical axis, you plot the person's score from a test that measures cardiovascular fitness (see the figure at the top of p. 115). The pattern of these data points suggests a line that slants from lower left to upper right, which is the opposite of the direction of slant in the first example. The figure at the top of p. 115 shows that the longer the person has owned the exercise machine, the better his or her cardiovascular fitness tends to be. This might be true in spite of the fact that time spent exercising decreases the longer the machine has been owned because purchasers of exercise machines might be starting from a point of low fitness, which may improve only gradually.

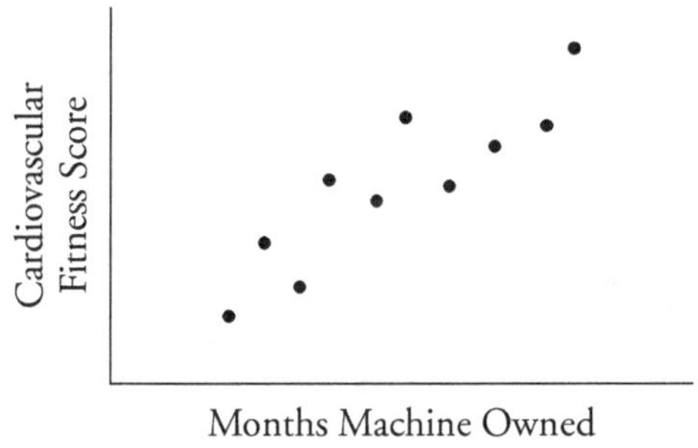

If two variables are positively correlated, as the value of one increases, so does the value of the other. If they are negatively (or inversely) correlated, as the value of one increases, the value of the other decreases.

A third possibility remains: that as the value of one variable increases, the value of the other neither increases nor decreases. The following figure is a scatter plot of months the exercise machine has been owned (horizontal axis) by the person's height (vertical axis). No line trend can be seen in the plot. New owners of exercise machines may be short or tall, and the same is true of people who have had their machines longer. These two variables appear to have no correlation.

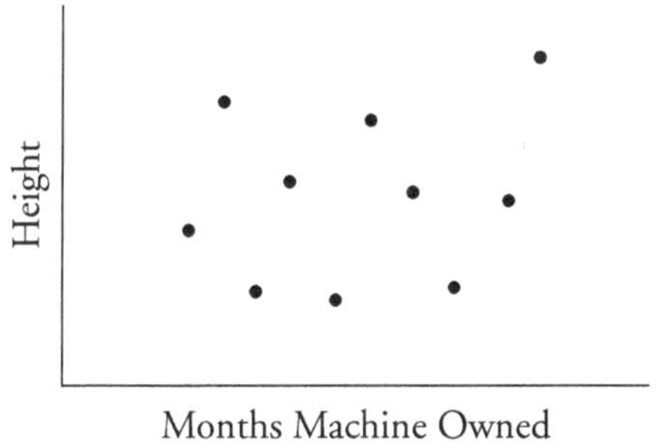

You can go even further in expressing the relationship between variables. Compare the two scatter plots in the figure below. Both plots show a positive correlation because, as the values on one axis increase, so do the values on the other. But the data points in plot (b) are more closely packed than the data points in plot (a), which are more spread out. If a line were drawn through the middle of the trend, the points in plot (b) would be closer to the line than the points in plot (a). In addition to direction (positive or negative), correlations can have strength, which is a reflection of the closeness of the data points to a perfect line. Plot (b) shows a stronger correlation than plot (a).

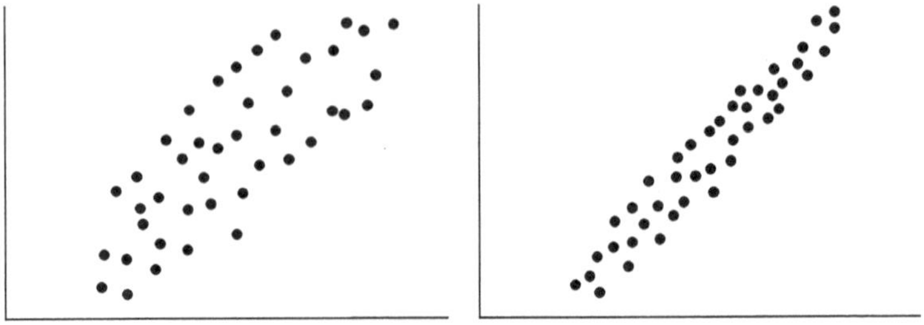

(a) (b)

Pearson's product moment coefficient (r), commonly referred to as the **correlation coefficient,** is a quantitative measure of correlation between two interval-level variables. The coefficient r can take values from -1.0 to 1.0. The sign of r indicates whether the correlation is positive or negative. The magnitude (absolute value) of r indicates the strength of the correlation, or how close the array of data points is to a straight line.

Two computing formulas for r are

$$r = \frac{\sum xy - \frac{1}{n}\left(\sum x\right)\left(\sum y\right)}{(n-1)s_x s_y} \quad \text{and} \quad r = \frac{\sum\left(\frac{x_i - \overline{x}}{s_x}\right)\left(\frac{y_i - \overline{y}}{s_y}\right)}{n-1}$$

where $\sum xy$ is the sum of the xy cross-products (each x multiplied by its paired y), n is the size of the sample (the number of data pairs), $\sum x$ and $\sum y$ are the sums of the x and y values, s_x and s_y are the sample standard deviations of x and y, and $\overline{x}$ and $\overline{y}$ are the means.

Example 1: Use Table 8-2 to compute r for the relationship between months of exercise-machine ownership and hours of exercise per week. The first step is to compute the components required in the main formula. Let x be months of ownership and y be hours of exercise, although you could also do the reverse.

Table 8-2 Determining the Correlation Coefficient for Table 8-1

x	y	xy	x^2	y^2
5	5	25	25	25
10	2	20	100	4
4	8	32	16	64
8	3	24	64	9
2	8	16	4	64
7	5	35	49	25
9	5	45	81	25
6	7	42	36	49
1	10	10	1	100
12	3	36	144	9
$\Sigma x = 64$	$\Sigma y = 56$	$\Sigma xy = 285$	$\Sigma x^2 = 520$	$\Sigma y^2 = 374$

$$\left(\sum x\right)^2 = 64^2 = 4{,}096$$

$$\left(\sum y\right)^2 = 56^2 = 3{,}136$$

Now, compute the sample standard deviations for x and y using the formula from Chapter 3 (pp. 40–41):

$$s_x = \sqrt{\frac{\sum x^2 - \frac{\left(\sum x\right)^2}{n}}{n-1}}$$

$$= \sqrt{\frac{520 - \frac{4{,}096}{10}}{10-1}}$$

$$= \sqrt{\frac{110.4}{9}}$$

$$= \sqrt{12.267}$$

$$= 3.502$$

$$s_y = \sqrt{\frac{374 - \frac{3,136}{10}}{10-1}}$$
$$= \sqrt{\frac{60.4}{9}}$$
$$= \sqrt{6.711}$$
$$= 2.591$$

Finally, r may be computed:

$$r_{x,y} = \frac{\sum xy - \frac{1}{n}(\sum x)(\sum y)}{(n-1)s_x s_y}$$
$$= \frac{285 - \frac{1}{10}(64)(56)}{(10-1)(3.502)(2.591)}$$
$$= \frac{-73.4}{81.663}$$
$$= -0.899$$

A correlation of $r = -0.899$ is almost as strong as the maximum negative correlation of -1.0, reflecting the fact that your data points fall relatively close to a straight line.

Finding the significance of *r*

You might want to know how significant an r of -0.899 is. The formula to test the null hypothesis that R (the population correlation) $= 0$ is

$$t = \frac{r\sqrt{(n-2)}}{\sqrt{(1-r^2)}}$$

where r is the sample correlation coefficient and n is the size of the sample (the number of data pairs). The probability of t may be looked up in Table B-3 in Appendix B using $n - 2$ degrees of freedom.

$$t = \frac{-0.899\sqrt{(10-2)}}{\sqrt{1-(-0.899)^2}} = \frac{-2.542}{0.438} = -5.804 \text{ (rounded to the nearest 1000th)}$$

The probability of obtaining a t of -5.804 with 8 degrees of freedom (drop the sign when looking up the value of t) is lower than the lowest listed probability of 0.0005. If the correlation between months of exercise-machine ownership and hours of exercise per week were actually 0, you would expect an r of -0.899 or lower in fewer than one out of a thousand random samples.

To evaluate a correlation coefficient, first determine its significance. If the probability that the coefficient resulted from chance is not acceptably low, the analysis should end there; neither the coefficient's sign nor its magnitude may reflect anything other than sampling error. If the coefficient is statistically significant, the sign should give an indication of the direction of the relationship; the magnitude indicates its strength. Remember, however, that all statistics become significant with a high enough n.

Even if it's statistically significant, whether a correlation of a given magnitude is substantively significant depends greatly on the phenomenon being studied. Generally, correlations tend to be higher in the physical sciences, where relationships between variables often obey uniform laws, and lower in the social sciences, where relationships may be harder to predict. A correlation of 0.4 between a pair of sociological variables may be more meaningful than a correlation of 0.7 between two variables in physics.

Bear in mind also that the correlation coefficient measures only straight-line relationships. Not all relationships between variables trace a straight line. The figure below shows a curvilinear relationship such that values of y increase along with values of x up to a point, then decrease with higher values of x. The correlation coefficient for this plot is 0, the same as for the plot in the second figure on p. 115. This plot, however, shows a relationship that the figure on p. 115 does not.

Correlation does not imply causation. The fact that Variable A and Variable B are correlated does not necessarily mean that A caused B or that B caused A (though either may be true). If you were to examine a database of demographic information, for example, you would find that the number of churches in a city is correlated with the number of violent crimes in the city. The reason is not that church attendance causes crime, but that these two variables both increase as a function of a third variable, population.

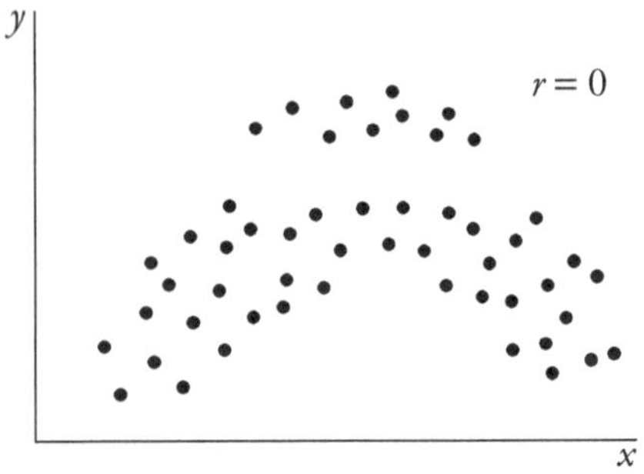

Also note that the scales used to measure two variables have no effect on their correlation. If you had converted hours of exercise per week to minutes per week, and/or months the machine was owned to days owned, the correlation coefficient would have been the same.

The coefficient of determination

The square of the correlation coefficient r is called the **coefficient of determination** and can be used as a measure of the proportion of variability that two variables share, or how much one can be "explained" by the other. The coefficient of determination for this example is $(-0.899)^2 = 0.808$. Approximately 80 percent of the variability of each variable in this example is shared with the other variable.

Simple Linear Regression

To describe the linear association between quantitative variables, a statistical procedure called **regression** often is used to construct a model. Regression is used to assess the contribution of one or more "explanatory" variables (called **independent** variables) to one "response" (or **dependent**) variable. It also can be used to predict the value of one variable based on the values of others. When there is only one independent variable and when the relationship can be expressed as a straight line, the procedure is called simple linear regression.

Any straight line in two-dimensional space can be represented by this equation:

$$y = a + bx$$

where y is the variable on the vertical axis, x is the variable on the horizontal axis, a is the y-value where the line crosses the vertical axis (often called the **intercept**), and b is the amount of change in y corresponding to a 1-unit increase in x (often called the **slope**). See figure below.

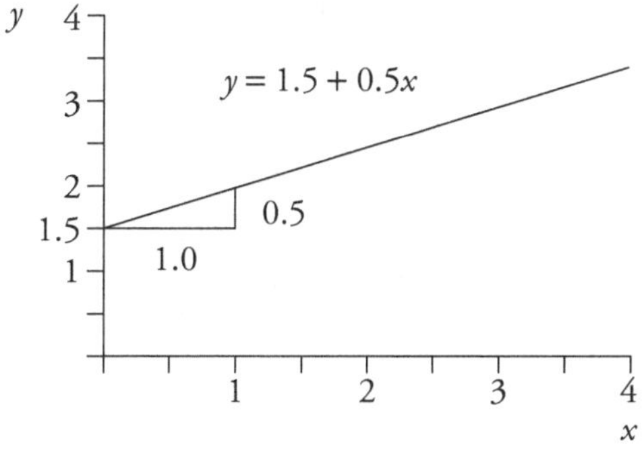

Returning to the exercise example, you observed that the scatter plot of points in the figure on p. 114 resembles a line. The regression procedure fits the best possible straight line to an array of data points. If no single line can be drawn such that all the points fall on it, what is the "best" line? Statisticians use the line that minimizes the sum of squared deviations from each data point to the line, a fact that will become clearer after you compute the line for the example.

Regression is an inferential procedure, meaning that it can be used to draw conclusions about populations based on samples randomly drawn from those populations. Suppose that your ten exercise-machine owners were randomly selected to represent the population of all exercise-machine owners. In order to use this sample to make educated guesses about the relationship between the two variables (months of machine ownership and time spent exercising) in the population, you need to rewrite the equation above to reflect the fact that you will be estimating population parameters:

$$y = \beta_0 + \beta_1 x$$

All you have done is replace the intercept (a) with β_0 and the slope (b) with β_1. The formula to compute the parameter estimate $\hat{\beta}_1$ is

$$\hat{\beta}_1 = \frac{S_{xy}}{S_{xx}}$$

where $S_{xy} = \sum xy - \frac{(\sum x)(\sum y)}{n}$ and $S_{xx} = \sum x^2 - \frac{(\sum x)^2}{n}$.

The formula to compute the parameter estimate $\hat{\beta}_0$ is

$$\hat{\beta}_0 = \bar{y} - \hat{\beta}_1 \bar{x}$$

where $\bar{y}$ and $\bar{x}$ are the two sample means.

You have already computed the quantities that you need to substitute into these formulas for the exercise example—except for the $\bar{x}$ mean of *x*, which is $\frac{64}{10} = 6.4$, and the $\bar{y}$ mean of *y*, which is $\frac{56}{10} = 5.6$. First, compute the estimate of the slope:

$$S_{xy} = 285 - \frac{(64)(56)}{10} = -73.4$$

$$S_{xx} = 520 - \frac{4,096}{10} = 110.4$$

$$\hat{\beta}_1 = \frac{-73.4}{110.4} = -.665$$

Now the intercept may be computed:

$$\begin{aligned}\hat{\beta}_0 &= \bar{y} - \hat{\beta}_1 \bar{x} \\ &= 5.6 - \left(-0.665(6.4)\right) \\ &= 5.6 - \left(-4.256\right) \\ &= 9.856\end{aligned}$$

So, the regression equation for the example is $\hat{y} = 9.856 - 0.665x$. When you plot this line over the data points, the result looks like that shown below.

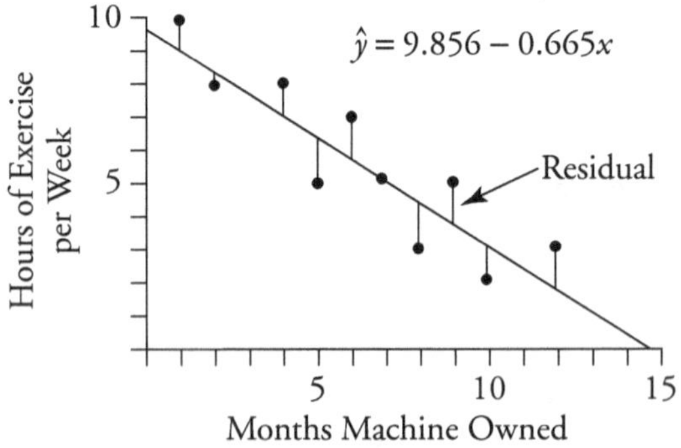

The vertical distance from each data point to the regression line is the error, or **residual,** of the line's accuracy in estimating that point. Some points have positive residuals (they lie above the line); some have negative residuals (they lie below the line). If all the points fell on the line, there would be no error and no residuals. The mean of the sample residuals is always 0 because the regression line is always drawn such that half of the error is above it and half below it. The equations that you used to estimate the intercept and slope determine a line of "best fit" by minimizing the sum of the squared residuals. This method of regression is called **least squares.**

Because regression estimates usually contain some error (that is, all points do not fall on the line), an error term (ε, the Greek letter epsilon) is usually added to the end of the equation:

$$y = \beta_0 + \beta_1 x + \varepsilon$$

The estimate of the slope β_1 for the exercise example was -0.665. The slope is negative because the line slants down from left to right, as it must for two variables that are negatively correlated, reflecting that one variable decreases as the other increases. When the correlation is positive, β_1 is positive, and the line slants up from left to right.

Confidence interval for the slope

What if the slope is 0, as in the figure below?

That means y has no linear dependence on x, or that knowing x does not contribute anything to your ability to predict y.

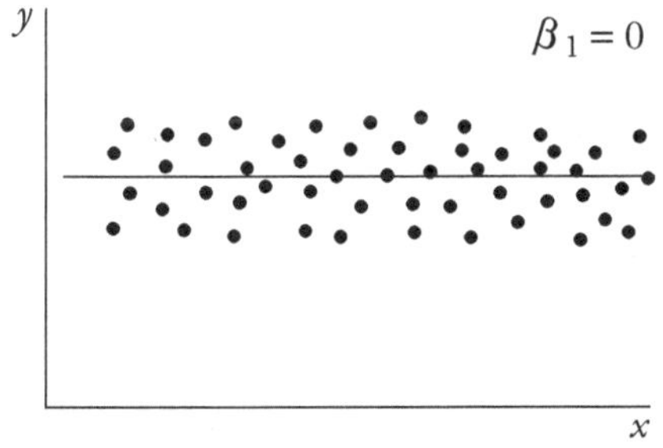

It is often useful to compute a confidence interval for a regression slope. If the confidence interval contained 0, you would be unable to conclude that x and y are related. The formula to compute a confidence interval for β_1 is

$$(a,\ b) = \hat{\beta}_1 \pm t_{a/2,\ n-2} \cdot \frac{s}{\sqrt{S_{xx}}}$$

where $s = \sqrt{\dfrac{\Sigma(y-\hat{y})^2}{n-2}}$ and $S_{xx} = \Sigma x^2 - \dfrac{(\Sigma x)^2}{n}$, and where $\Sigma(y-\hat{y})^2$ is the sum of the squared residuals, $t_{\alpha/2,\ n-2}$ is the critical value from the t-table corresponding to half the desired alpha level at $n - 2$ degrees of freedom, and n is the size of the sample (the number of data pairs). The test for this example will use an alpha of 0.05. Table B-3 in Appendix B shows that $t_{.025,\ 8} = 2.306$.

Compute the quantity $\Sigma(y-\hat{y})^2$ by subtracting each predicted y-value $(\hat{y})$ from each actual y-value, squaring it, and summing the squares (see Table 8-3). The predicted y-value $(\hat{y})$ is the y-value that would be predicted from each given x, using the formula $\hat{y} = 9.856 - 0.665x$.

Table 8-3 Determining the Residuals for the Data in Table 8-1

x	y		$\hat{y}$		*residual*	*residual*^{2}
5	5	–	6.531	=	–1.531	2.344
10	2	–	3.206	=	–1.206	1.454
4	8	–	7.196	=	0.804	0.646
8	3	–	4.536	=	–1.536	2.359
2	8	–	8.526	=	–0.526	0.277
7	5	–	5.201	=	–0.201	0.040
9	5	–	3.871	=	1.129	1.275
6	7	–	5.866	=	1.134	1.286
1	10	–	9.191	=	0.809	0.654
12	3	–	1.876	=	1.124	1.263
					0	11.600

Now, compute s:

$$s = \sqrt{\frac{11.600}{10-2}} = \sqrt{1.45} = 1.204$$

You have already determined that $S_{xx} = 110.4$ (p. 122), so you can proceed to the main formula:

$$\begin{aligned}(a,\ b) &= -0.665 \pm 2.306\frac{1.204}{\sqrt{110.4}} \\ &= -0.665 \pm 2.306\left(\frac{1.204}{10.507}\right) \\ &= -0.665 \pm 2.306(0.1146) \\ &= -0.665 \pm 0.264 \\ &= (-0.929,\ -0.401)\end{aligned}$$

You can be 95 percent confident that the population parameter $\hat{\beta}_1$ (the slope) is no lower than −0.93 and no higher than −0.40 (rounded to the nearest hundredth). Because this interval does not contain 0, you would be able to reject the null hypothesis that $\hat{\beta}_1 = 0$ and conclude that these two variables are indeed related in the population.

Confidence interval for prediction

You have learned that you could predict a *y*-value from a given *x*-value. Because there is some error associated with your prediction, however, you might want to produce a confidence interval rather than a simple point estimate. The formula for a prediction interval for *y* for a given *x* is

$$(a,\ b) = \hat{y} \pm t_{a/2,\ n-1}\left(s\sqrt{1+\frac{1}{n}+\frac{(x-\overline{x})^2}{S_{xx}}}\right)$$

where $S_{xx} = \sum x^2 - \frac{(\sum x)^2}{n}$ and $s = \sqrt{\frac{\Sigma(y-\hat{y})^2}{n-2}}$, and where $\hat{y}$ is the *y*-value predicted for *x* using the regression equation, $t_{\alpha/2,\ n-2}$ is the critical value from the *t*-table corresponding to half the desired alpha level at $n - 2$ degrees of freedom, and *n* is the size of the sample (the number of data pairs).

Example 2: What is a 90 percent confidence interval for the number of hours spent exercising per week if the exercise machine is owned 11 months?

The first step is to use the original regression equation to compute a point estimate for *y:*

$$\begin{aligned}\hat{y} &= 9.856 - 0.665(11) \\ &= 9.856 - 7.315 \\ &= 2.541\end{aligned}$$

For a 90 percent confidence interval, you need to use $t_{.05,\ 8}$, which Table B-3 in Appendix B shows to be 1.860. You have already computed the remaining quantities, so you can proceed with the formula (round your answers to the nearest thousandth):

$$\begin{aligned}(a,\ b) &= 2.541 \pm 1.860(1.204)\sqrt{1+\frac{1}{10}+\frac{(11-6.4)^2}{110.40}} \\ &= 2.541 \pm 1.860(1.204)\sqrt{1+0.10+\frac{21.16}{110.40}} \\ &= 2.541 \pm 1.860(1.204)\sqrt{1.292} \\ &= 2.541 \pm 2.546 \\ &= (-0.005,\ 5.087)\end{aligned}$$

You can be 90 percent confident that the population mean for the number of hours spent exercising per week when x (number of weeks machine owned) = 11 is between about 0 and 5.

Assumptions and cautions

The use of regression for parametric inference assumes that the errors (ε) are (1) independent of each other, and (2) normally distributed with the same variance for each level of the independent variable. The following figure shows a violation of the second assumption. The errors (residuals) are greater for higher values of x than for lower values.

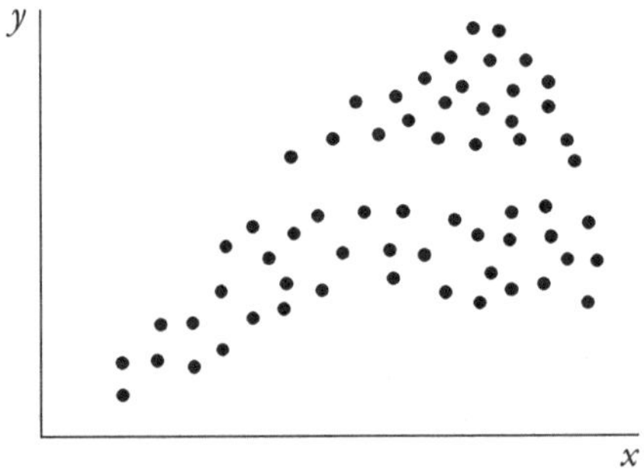

Least squares regression is sensitive to **outliers,** or data points that fall far from most other points. If you were to add the single data point $x = 15$, $y = 12$ to the exercise data, the regression line would change to the dotted line shown in the figure at the top of the next page. You need to be wary of outliers because they can greatly influence the regression equation.

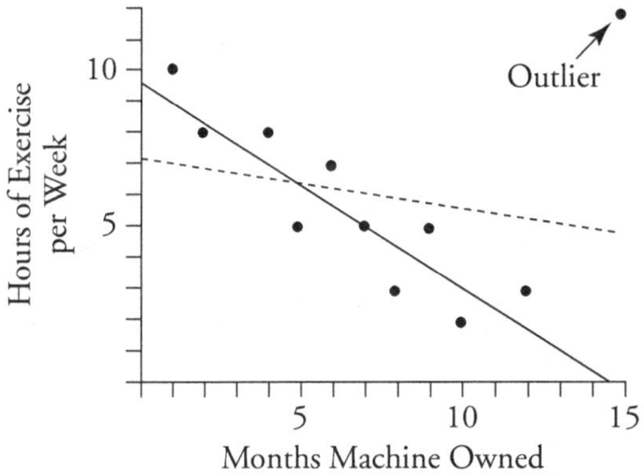

It can be dangerous to **extrapolate** in regression—to predict values beyond the range of your data set. The regression model assumes that the straight line extends to infinity in both directions, which often is not true. According to the regression equation for the example, people who have owned their exercise machines longer than around 15 months do not exercise at all. It is more likely, however, that "hours of exercise" reaches some minimum threshold and then declines only gradually, if at all (see the following figure).

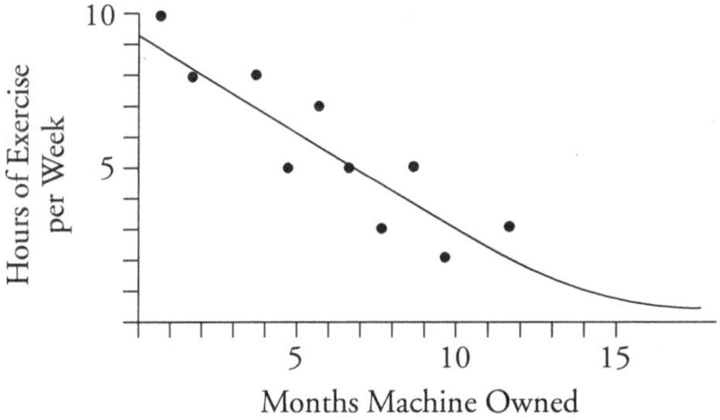

Causation and correlation

Causation is defined as a chain of events that occurs between two variables so that the first variable results in an action in the second variable. The causal relationship between the two variables involves *time*. In other words, the second event always follows the occurrence of the first event.

It is a cause-and-effect relationship. The first event is the *cause* and the second event is the *effect*. For example, when the CEO of a well-known technology company died, the effect was that the stock market price of the company suddenly declined the day after his death.

Correlation is a statistical method used to describe the *direction* and *magnitude* of two variables. It is important to learn how to distinguish between correlation and causation because *correlation does not always imply causation*. For example, evidence from medical research frequently states that people who smoke cigarettes (on average) decrease their lifespan. This statement may tempt people to jump to the conclusion that smoking cigarettes *causes* people to have a shorter lifespan. However, statistically we should be careful about drawing conclusions that may not be 100 percent verifiable. In this example, compounding factors may cause people to have a shorter lifespan (genetics, stress, diet, lifestyle, trauma, etc.). A statistical investigation with supporting relevant data helps us to draw relevant conclusions about whether or not smoking is the real cause of a shorter life, or if smoking is only correlated with shorter life.

Chi-Square (χ^2)

The statistical procedures that we have reviewed thus far are appropriate only for numerical variables. The **chi-square** (χ^2) **goodness of fit test** can be used to evaluate a relationship between two categorical variables. With the help of the **chi-square distribution,** a confidence interval that estimates a population's standard deviation can be constructed. The sample statistic chi-squared has a sampling chi-squared distribution, which has only positive values, or zero. The values of chi-square distribution change depending on the associated degree of freedom. As the number of degrees of freedom increases, the chi-square distribution approaches a normal distribution.

The chi-square is one example of a **nonparametric test.** Nonparametric tests are used when assumptions about normal distribution in the population cannot be met. These tests are less powerful than parametric tests.

Suppose that 125 children are shown three television commercials for breakfast cereal and are asked to pick which commercial they liked best. The results are shown in Table 8-4.

You would like to know if the choice of favorite commercial is related to whether the child is a boy or a girl or if these two variables are independent. The totals in the margins will allow you to determine the overall probability of (1) liking commercial A, B, or C, regardless of gender; and (2) being either a boy or a girl, regardless of favorite commercial. If the two variables are independent, then you should be able to use these probabilities to predict approximately how many children should be in each cell. If the actual count is very different from the count that you would expect if the probabilities are independent, the two variables must be related.

Table 8-4 Commercial Preference for Boys and Girls

	A	*B*	*C*	*Totals*
Boys	30	29	16	75
Girls	12	33	5	50
Totals	42	62	21	125

Consider the upper-right cell of the table. The overall probability of a child in the sample being a boy is $75 \div 125 = 0.6$. The overall probability of liking Commercial A is $42 \div 125 = 0.336$. The multiplication rule (covered in Chapter 4) states that the probability of both of two independent events occurring is the product of their two probabilities. Therefore, the probability of a child both being a boy and liking Commercial A is $0.6 \times 0.336 = 0.202$. The expected number of children in this cell, then, is $0.202 \times 125 = 25.25$.

There is a faster way of computing the expected count for each cell: Multiply the row total by the column total and divide by *n*. The expected count for the first cell is, therefore, $(75 \times 42) \div 125 = 25.25$. If you perform this operation for each cell, you get the expected counts (in parentheses) shown in Table 8-5.

Table 8-5 Chi-Square Results for Table 8-4

	A	*B*	*C*	*Totals*
Boys	30 (25.25)	29 (37.2)	16 (12.6)	75
Girls	12 (16.8)	33 (24.8)	5 (8.4)	50
Totals	42	62	21	125

Note that the expected counts properly add up to the row and column totals. You are now ready for the formula for χ^2, which compares each cell's actual count to its expected count:

$$\chi^2 = \Sigma \frac{(\text{observed} - \text{expected})^2}{\text{expected}}$$

The formula describes an operation that is performed on each cell and which yields a number. When all the numbers are summed, the result is χ^2. Now, compute it for the six cells in the example:

$$\begin{aligned}\chi^2 &= \frac{(30-25.2)^2}{25.2} + \frac{(29-37.2)^2}{37.2} + \frac{(16-12.6)^2}{12.6} + \frac{(12-16.8)^2}{16.8} \\ &\quad + \frac{(33-24.8)^2}{24.8} + \frac{(5-8.4)^2}{8.4} \\ &= 0.914 + 1.808 + 0.917 + 1.371 + 2.711 + 1.376 \\ &= 9.097\end{aligned}$$

The larger χ^2, the more likely that the variables are related: Note that the cells that contribute the most to the resulting statistic are those in which the expected count is very different from the actual count.

Chi-square has a probability distribution, the critical values for which are listed in Table B-4 in Appendix B. As with the *t*-distribution, χ^2 has a degrees-of-freedom parameter, the formula for which is

$$(\text{number of rows} - 1) \times (\text{number of columns} - 1)$$

or in your example:

$$(2 - 1) \times (3 - 1) = 1 \times 2 = 2$$

In Table B-4 in Appendix B, a chi-square of 9.097 with two degrees of freedom falls between the commonly used significance levels of 0.02 and 0.01. If you had specified an alpha of 0.05 for the test, you could, therefore, reject the null hypothesis that gender and favorite commercial are independent. At $a = 0.01$, however, you could not reject the null hypothesis.

The χ^2 test does not allow you to conclude anything more specific than that there is some relationship in your sample between gender and commercial liked (at $\alpha = 0.05$). Examining the observed versus expected counts in each cell might give you a clue as to the nature of the relationship and which levels of the variables are involved. For example, Commercial B appears to have been liked more by girls than boys. But χ^2 tests only the very general null hypothesis that the two variables are independent.

Sometimes a chi-square test of homogeneity of populations is used. It is very similar to the test for independence. In fact, the mechanics of these tests are identical. The real difference is in the design of the study and the sampling method.

Chapter Check-Out

Questions

For problems 1 through 3, consider this data on the dose of Medicine X administered and the length of time of continued illness.

Dose	0	1	2	3	4	5	6
Time to cure (days)	15.4	12.2	13.7	9.2	9.9	6.1	4.1

1. **(a)** Does a scatter plot of the data show negative or positive correlation?

(b) Calculate the correlation coefficient.

(c) Find the significance of the correlation coefficient.

2. **(a)** Apply linear regression to the data to find the slope and intercept.

(b) Calculate the 95 percent confidence interval for the slope of your regression.

3. **(a)** Predict the result if the dose was 1.5, and give the 90 percent confidence interval for your prediction.

(b) Predict the result if the dose was 8, and give the 90 percent confidence interval for your prediction.

4. Which of following five choices describe a correlation? (Select **all** that apply.)
   - **A.** Insurance companies charge male drivers more than female drivers. The rate of the accidents is higher for male drivers.
   - **B.** The relationship between the length of the palm of your hand and how long you will live (number of years)
   - **C.** The amount of carbonated soda that an adult between 30 to 60 years old drinks, and his or her likelihood of obesity
   - **D.** Data results that confirm the duration of the influenza virus among adults is significantly lengthier in the winter for those people residing on the east coast
   - **E.** Data results that provide evidence that people who exercise for at least 30 minutes a day burn more calories within 24 hours after they exercise

Answers

1. **(a)** negative; **(b)** -0.9511; **(c)** $t = -6.885, p < 0.0005$
2. **(a)** $\beta_1 = -1.78$; $\beta_0 = 15.4321$; **(b)** -1.78 ± 0.66
3. **(a)** 12.76 ± 3.05; **(b)** 1.19 ± 3.94
4. A, B, and C

REVIEW QUESTIONS

Use this Review to practice what you've learned in this book. After you work through the review questions, you'll be well on your way to achieving your goal of understanding basic statistical methods.

Questions

Chapter 2

1. What is wrong with this pie chart?

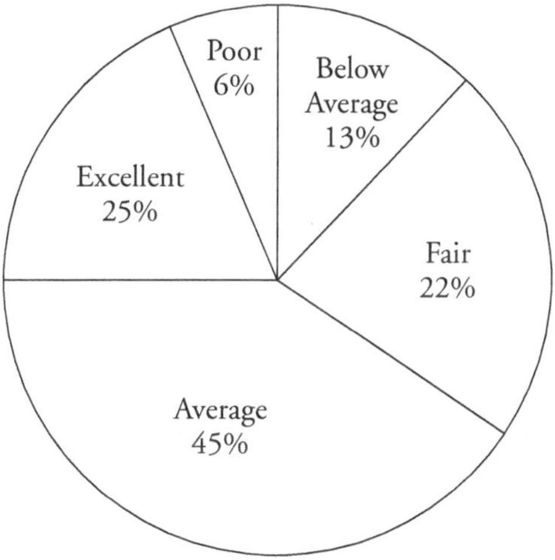

2. Study the following frequency histogram of the inventory at a hat store. The total number of hats is 1,000. Then answer these questions:

(a) How many hats are blue?

(b) How many hats are white or black?

(c) If a hat is chosen randomly, what is the chance that it is multicolored?

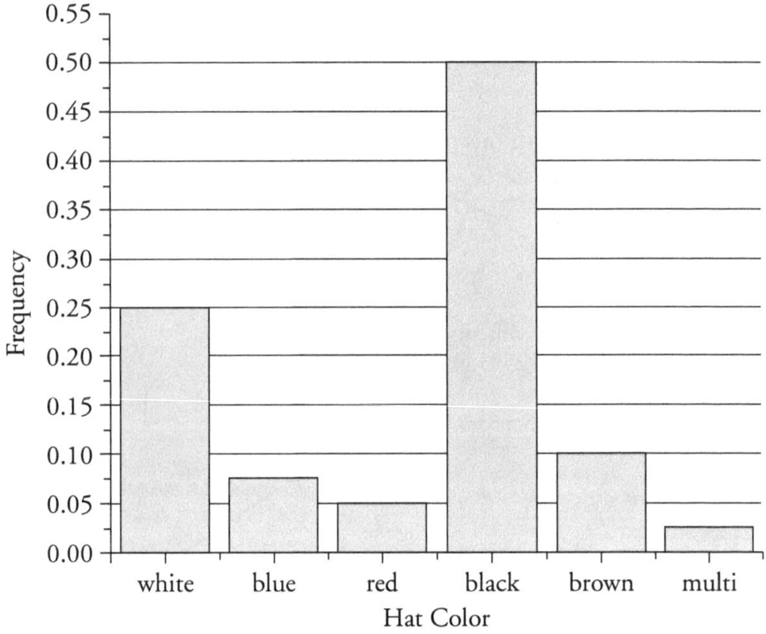

Chapter 3

3. True or False: *Median* and *mean* are two terms for the same quantity.

4. True or False: *Variance* and *standard deviation* are two terms for the same quantity.

Questions 5 and 6 refer to the following data set collected and sorted for a study conducted by the regional cancer institute. The data was collected from 14 participants in May through September. The data represents the diminished size (in millimeters) of each participant's polyp after 5 months of treatment.

1.0, 1.2, 1.8, 5.8, 6.1, 6.7, 6.7, 6.7, 6.8, 6.9, 7, 7.1, 7.1, 9.0

5. Find the range, midrange, and standard deviation of the data set.

6. Based on the data set, which of the following are sensitive to outliers?

Select **all** that apply.

A. range

B. midrange

C. standard deviation

Chapter 4

7. True or False: The binomial distribution applies only to events that have two outcomes: "success" or "failure."

8. A fair die, with six sides, is tossed twice. The chance of rolling two sixes is

A. $\frac{2}{6}$

B. $\frac{1}{6}$

C. $\frac{1}{12}$

D. $\frac{1}{36}$

E. $\frac{1}{4}$

9. Two fair dice, with six sides each, are rolled. The probability that the sum of the dice equals 12 is

A. 0

B. 1

C. $\frac{1}{6}$

D. $\frac{1}{12}$

E. $\frac{1}{36}$

10. Two fair dice, with six sides each, are rolled. The probability that the sum of the dice equals 7 is

A. 0

B. 1

C. $\frac{1}{6}$

D. $\frac{1}{12}$

E. $\frac{1}{36}$

11. A population with a normal distribution has a mean of 1 and a variance of 4. The most information you can provide about the probability of randomly choosing a value greater than 7 from this sample is that it is

A. 0

B. less than 0.15 percent

C. less than 0.3 percent

D. less than 32 percent

E. less than 16 percent

12. A slot machine has four windows, and each can contain one of eight different symbols. To win the jackpot, all four windows must contain the same symbol. The chance of winning a jackpot is

A. 0

B. $\frac{1}{8}$

C. $\left(\frac{1}{8}\right)^3$

D. $\frac{1}{4}$

E. 1

13. The probability that a high school student receives acceptance to a prestigious university is 20%. What is the probability that 6 out of 12 students who applied this year will receive admission to a prestigious university?

A. 1.6%

B. 15%

C. 85%

D. 8.5%

E. 67%

Chapter 5

14. True or False: In a normally distributed population, the standard deviation may never be larger than the mean.

15. True or False: In this instance, μ and $\bar{x}$ are the same thing.

16. Which of the following is NOT a correct null hypothesis H_0?

A. $\mu = 1$

B. $\mu_1 = \mu_2$

C. $\pi = 0.5$

D. $\pi_1 = \pi_2$

E. $\bar{x} = 1$

Questions 17 and 18 refer to the following graph.

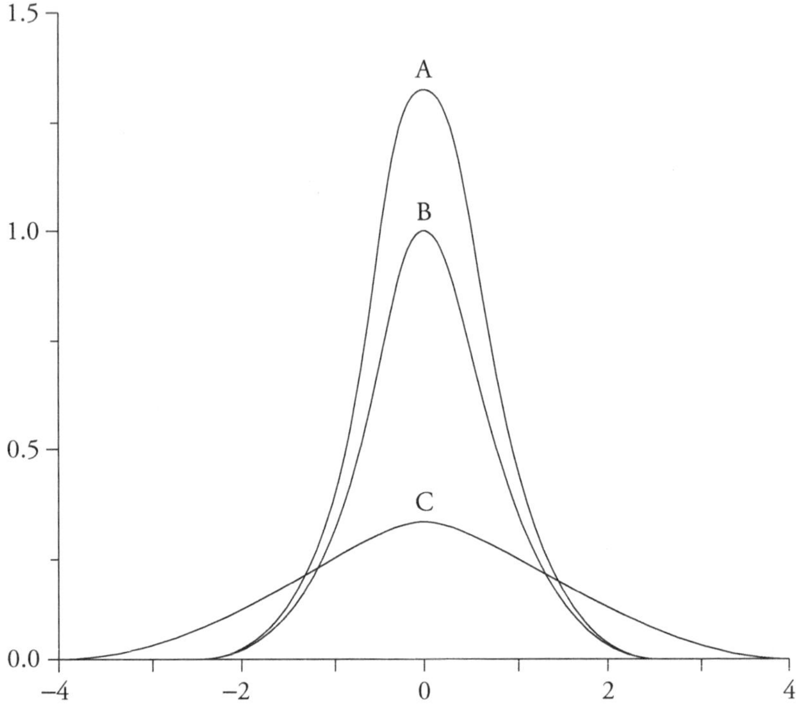

17. Which normal curve pictured above has the greatest standard deviation?

18. Which normal curve in the graph above has the smallest mean?

Chapter 6

19. True or False: If two variables are correlated, then they must be causally related.

20. Rejecting the null hypothesis when it is actually true is

A. not an error

B. a Type I error

C. a Type II error

D. neither a Type I nor a Type II error

E. impossible

Chapter 7

21. A machine for filling beer bottles fills the bottles so that the distribution has a mean of 12 ounces of beer in each bottle, with a standard deviation of 0.05 ounce. What is the probability that a randomly selected bottle contains at least 11.9 ounces?

22. Suppose that there are only two candidates (A and B) for president and that in a particular state, 5,000,000 votes will be cast.

(a) What percentage of the total votes should a TV station count to be sure to be within ± 1 percent that Candidate A has won at a 95 percent confidence level?

(b) The first 20,000 votes are rapidly counted. Candidate A has received 52 percent of the votes, and Candidate B has received 48 percent of the votes. What is the width of the 99.9 percent confidence interval on Candidate A's result?

23. At a major airport, a researcher studies the average length of delays (excluding flights that leave on time). His data is shown below, for ten randomly chosen days.

Day	1	2	3	4	5	6	7	8	9	10
Average delay (hours)	2.0	1.8	0.32	0.41	2.3	0.72	0.22	0.28	0.57	3.2

(a) Calculate the mean and standard deviation of the delays, as well as the median.

(b) What is the 95 percent confidence interval on the average length of delay?

(c) Compare the mean and median from part (a). What is the p-value for a delay of the median size?

(d) Do you think this population is normally distributed?

Chapter 8

For questions 24–26, evaluate each relationship below and determine if the variables show correlation or causation.

24. A windmill rotates faster when the wind is strong.

25. Children who do not get enough sleep are much more likely to experience migraine headaches.

26. There is a positive relationship between eating breakfast daily and obesity.

27. True or False: If the correlation coefficient for a data set is 0, then there is no relationship between the dependent and independent variables.

28. True or False: The χ^2 test tells you only if two variables are independent.

29. True or False: The correlation coefficient is equal to coefficient of determination.

30. True or False: The correlation coefficient ranges from −2 to 2.

31. A microbiologist is testing a new antibiotic she developed against a common bacterial infection. She applies each drug to different plates of bacterial cultures and measures the fraction of bacteria that dies after 1 day. Her results are given in the following table.

Dose of drug (mg)	0	0.5	1.0	1.5	2.0	2.5	3.0
New drug (% dead)	2.0	12.3	27.1	32.2	48.9	56.6	60.6
Common drug (% dead)	3.1	6.0	16.5	12.3	25.9	29.0	42.5

Using the information in the preceding table, answer these questions:

(a) Calculate the coefficient of correlation for each drug.

(b) What is the significance level of each of these correlations?

(c) Calculate the slope for each plot.

(d) Do the 95 percent confidence intervals for the two slopes overlap?

32. Consider the following data on the color of a dog and the average annual income of the dog's owner. Compute a χ^2 value for this data and determine the *p*-level to which these variables are related.

	$0–$25,000	***$25,001–$50,000***	***$50,001–$75,000***	***$75,000+***	***Totals***
White	10	12	17	23	62
Black	15	28	13	11	67
Orange/golden	7	9	11	15	42
Mixed	24	26	32	42	124
Total	56	75	73	91	295

Answers

1. More than 100 percent is shown.
2. **(a)** 75; **(b)** 750; **(c)** 0.025
3. False
4. False
5. range: 8.0; midrange: 5; standard deviation: 2.48
6. A (range) and B (midrange) are both sensitive to outliers.
7. True
8. D
9. E
10. C
11. B
12. C
13. A
14. False
15. False
16. E
17. curve C
18. None—they are all equal.
19. False
20. B
21. 97.72 percent
22. **(a)** 0.192 percent; **(b)** 2.3 percent
23. **(a)** mean: 1.18; standard deviation: 1.06; median: 0.65; **(b)** (0.42, 1.94); **(c)** $0.05 < p < 0.1$; **(d)** no
24. correlation
25. causation
26. correlation
27. False

28. True

29. False

30. False

31. **(a)** new drug: 0.989853; common drug: 0.9582897; **(b)** new drug: $p < 0.0005$; common drug: $p < 0.0005$; **(c)** new drug: 20.44; common drug: 12.4; **(d)** no

32. $\chi^2 = 17.03406$; $0.025 < p < 0.05$

GLOSSARY

addition rule: For mutually exclusive random events, the chance of at least one of them occurring is the sum of their individual probabilities.

alternative hypothesis: An experimental hypothesis. It is the hypothesis that is supported if the null hypothesis is rejected. In math, the alternative hypothesis contains a statement of inequality (≠). The symbol for the alternative hypothesis is H_a or H_1. An example of H_a is when an experimental drug treatment study shows a measureable change in the participants. Also called a *research hypothesis.*

bar chart: A graphic that displays how data fall into different categories or groups.

Bayes' theorem (Bayes' law or **Bayes' rule):** Bayes' theorem describes the relationship between *conditional probability* and its reverse. The formula is

$$P(A \mid B) = \frac{P(B \mid A) \cdot P(A)}{P(B)}$$

bell-shaped curve: Symmetrical, single-peaked frequency distribution. Also called the *normal curve* or *Gaussian curve* or *mound-shaped curve.*

bias: The systematic underestimation or overestimation of a population parameter because of the method used to select the sample or the method used to obtain the information. See also *systematic error.*

bimodal curve: Curve with two equal scores of highest frequency.

binomial: An event with only two possible outcomes.

binomial probability distribution: For binomial events, the frequency of the number of favorable outcomes. For a large number of trials, the binomial distribution approaches the normal distribution.

bivariate: Involving two variables; when attempting to show a correlation between two variables, the analysis is said to be bivariate.

box plot (box-and-whisker plot): A graphic display of data indicating symmetry and central tendency.

categorical variable: A variable that assigns each object of interest to one of the available categories.

causation: A causation relationship indicates that one event is the result of the occurrence of the other event.

central limit theorem: A rule that states that regardless of the population (even a strong non-normal population), the sampling distribution of the mean from any population will be approximately normal for large sample sizes (over 30).

chi-square distribution: A probability distribution used to test the independence of two categorical variables.

chi-square goodness-of-fit test: An inferential test that shows whether or not a frequency distribution fits as expected or claimed.

class frequency: The number of observations that fall into a class interval.

class intervals: Categories or groups contained in frequency graphics.

coefficient of determination: A measure of the proportion of each other's variability that two variables share.

conditional probability: The measure of the probability of an event A is conditional on knowing that another event B has already occurred.

confidence interval (CI): The range of values that a population parameter could take at a given level of significance; the estimated ± margin of error.

confidence level: A selected percentage of confidence for a confidence interval.

continuous variable: A variable that can be measured with whole numbers and fractional (or decimal) parts thereof.

correlation (correlational relationship): Describes the statistical relationship between two variables. For example, two or more quantities (variables) correlate when they change together in a consistent manner. Thus, if the value of one variable is known, the other can be predicted from their relationship. However, correlation does *not* imply causation because there may be an unknown factor that influences both variables.

correlation coefficient: A measure of the degree to which two variables are linearly related. If one set of values increases and the other set tends to increase, it is called a *positive relationship*. Also called a *Pearson's product moment coefficient*.

critical value: The value of a computed statistic used as a threshold to decide whether the null hypothesis will be rejected. Also called *tabled value*.

data: Numerical information about variables; the measurements or observations to be analyzed with statistical methods.

degrees of freedom: A parameter used to help select the critical value in some probability distributions.

dependent events: Events such that the outcome of one has an effect on the probability of the outcome of the other.

dependent variable: A variable that is caused or influenced by another.

descriptive statistics: A branch of statistics in which numerical data are used to describe phenomena.

deviation: The distance of a value in a population (or sample) from the mean value of the population (or sample).

directional test: A test of the prediction that one value is higher than another; also called a *one-tailed test*.

discrete variable: A variable that can be measured by means of whole numbers only, or one that assumes only a certain set of definite values and no others.

disjoint occurrence: Both outcomes unable to happen at the same time.

distribution: A collection of measurements; how scores tend to be dispersed about a measurement scale.

dot plot: A graphic chart of data points that displays the measure of a small set of values.

double-counting: A mistake encountered in calculating the probability of at least one of several events occurring, when the events are not mutually exclusive. In this case, the addition rule does not apply.

empirical rule: A rule that describes the percentage of values that fall within intervals for a normal distribution. Sometimes called the *68-95-99.7 rule.*

extrapolate: To estimate the future values of a certain variable based on what has been observed.

frequency distribution: The frequency of occurrence of the values of a variable. For each possible value of the variable, there is an associated frequency at which the variable assumes that value.

frequency histogram: A graphic that displays how many measures fall into different classes, giving the frequency at which each category is observed.

frequency polygon: A graphic presentation of frequency of a phenomenon that typically uses straight lines and points.

grouped data: Data that has been sorted into intervals, usually in order to construct a frequency table.

grouped measures: A set of values that belong to the same class.

histogram: A vertical or horizontal bar graphic presentation of the frequency of a numerical variables.

hypothesis: A theoretical supposition about the characteristics of a population as a starting point for further research.

independent events: Events such that the outcome of one has no effect on the probability of the outcome of the other.

independent variable: A variable that causes or influences another variable.

inference: An educated statistical guess about a population parameter based upon analysis of a sample statistic. Inferences are always stated with a confidence or significance level.

inferential statistics: A branch of statistics that aims to draw conclusions about a population from the observed characteristics of the sample data.

intercept: There are x intercepts and y intercepts. An x intercept is the value of x at which a line crosses the horizontal axis. A y intercept is the value of y at which a line crosses the vertical axis.

interquartile range (IQR): The difference between the upper quartile (75th percentile) and the lower quartile (25th percentile).

interval scale: A scale that measures by using equal intervals to compare differences between pairs of values, such as the Fahrenheit temperature scale, measured in degrees. The difference between two values is meaningful.

J-shaped curve: A graph that pictorially shows how the exponential growth rate increases rapidly in the beginning, but eventually slows down and becomes constant. It displays an asymmetrical skewed distribution.

joint occurrence: Two or more independent events happening simultaneously; $P(A \cap B)$.

least squares: Any line- or curve-fitting model that minimizes the squared distance of data points to the line.

lower quartile: (Q_1), the 25th percentile of a set of measures.

margin of error: The margin of error measures the standard amount of the difference between a sample mean and the population mean. It is an acceptable probability of being wrong in testing a null hypothesis which makes the test results more reliable.

mean: The sum of the measures in a distribution divided by the number of measures; the average.

measures of central tendency: Descriptive measures that indicate the center of a set of values—for example, mean, median, and mode.

measures of variation: Descriptive measures that indicate the dispersion of a set of values—for example, variance, standard deviation, and standard error of the mean. Also called *spread.*

median: The middle measure in an ordered distribution with an odd number of values or the average of the two middle scores with an even number of values.

middle quartile: (Q_2), the 50th percentile of a set of measures; the median.

midrange: The sum of the largest and smallest values divided by two.

mode: Most frequent measure in a distribution; the high point on a frequency distribution.

mound-shaped curve: See *bell-shaped curve.*

multiplication rule: States that the probability of two or more independent events (hence, not mutually exclusive events) all occurring is the product of their individual probabilities.

mutually exclusive: Events such that the occurrence of one precludes the occurrence of the other. See also *disjoint occurrence.*

negative relationship: A relationship between two variables such that when one increases, the other decreases.

negatively skewed curve: A probability or frequency distribution that is not normal but rather is shifted such that the mean is less than the mode.

nominal scale: A scale using numbers, symbols, or names to designate different subclasses.

nondirectional test: A test of the prediction that two values are equal or a test that they are not equal. Also called a *two-tailed test.*

nonparametric test: A statistical test used when the population distribution is not normal or does not need to be characterized.

normal distribution: A specific continuous probability distribution with a graph that is a smooth bell-shaped curve symmetrical about the mean. Also called a *normal curve.*

null hypothesis: The reverse of the research hypothesis. The null hypothesis is directly tested by statistical analysis so that it is either rejected or not rejected, with a significance level. If the null hypothesis is rejected, the alternative hypothesis is supported. In math, the null hypothesis contains the statement of equality (=). The symbol for null hypothesis is H_0.

numerical statistics: Statistical parameters presented as numbers (as opposed to pictorial statistics).

ogive: A cumulative line graph that displays a running total from a frequency table. The upper value of any point on the graph shows what all of the frequencies up to that point would be when added together.

one-tailed test: A test of the prediction that one value is higher than another. Also called a *directional test.*

ordinal scale: A scale using names or symbols to rank order; its intervals are unspecified.

outlier: A data point that falls far from most other points; a score extremely divergent from the other measures of a set.

parameter: A characteristic of a population. The goal of statistical analysis is usually to estimate population parameters using statistics from a sample of the population.

Pearson's product moment coefficient: Quantitatively measures the degree and direction of the linear relationship between two variables. Also called a *correlation coefficient.*

percentile: The value in an ordered set of measurements such that P percent of the measures lie below that value.

pictorial statistics: Statistical parameters that are presented as graphs or charts (as opposed to simply as numbers).

pie chart: A graphic that displays parts of the whole in the form of a circle with its area divided proportionally.

point estimate: A number computed from a sample to represent a population parameter.

population: A group of phenomena that have something in common. The population is the larger group, whose properties (parameters) are estimated by taking a smaller sample from within the population and applying statistical analysis to the sample.

positive relationship: A relationship between two variables such that when one increases, the other increases, or when one decreases, the other decreases.

positively skewed curve: A probability or frequency distribution that is not normal, but rather is shifted such that the mean is greater than the mode.

power: The probability that a test will reject the null hypothesis when it is, in fact, false.

probability: A quantitative measure of the chances for a particular outcome or outcomes.

probability distribution: A function that determines the probability of each value of a discrete random variable.

proportion: For a binomial random event, the probability of a successful (or favorable) outcome in a single trial.

qualitative data (qualitative variable): Data that can be measured in categories, such as eye color or ice cream flavor.

quantitative data (quantitative variable): Data measurements or information about quantities from a study that can be described with numbers, such as size, temperature, and so on.

random error: An error that occurs as a result of sampling variability, through no direct fault of the sampler. It is a reflection of the fact that the sample is smaller than the population; for larger samples, the random error is smaller.

random sample: A sample selected from the population that guarantees each member of the population has the same chance of being selected into the sample.

range: The difference between the largest and smallest measures of a set.

ratio scale: A scale using numbers to rank order; its intervals are equal and the scale has an absolute-zero point.

region of acceptance: The area of a probability curve in which a computed test statistic will lead to acceptance of the null hypothesis.

region of rejection: The area of a probability curve in which a computed test statistic will lead to rejection of the null hypothesis.

regression: A statistical procedure used to estimate the linear dependence of one or more independent variables on a dependent variable.

relative frequency: The ratio of class frequency to total number of measures.

relative frequency histogram: A bar graph with columns that compare the proportion of data values of particular intervals.

relative frequency theory of probability: If a random event is repeated a large number of times, then the proportion of times that a particular outcome occurs is close to the probability of that outcome occurring in a single event.

research hypothesis: A prediction or expectation to be tested. If the null hypothesis is rejected, then the research hypothesis (also called the *alternative hypothesis*) is supported.

residual: The difference of an actual value y and its predicted value; a residual is an error.

sample: A group of members of a population selected to represent that population. A sample to which statistical analysis is applied should be randomly drawn from the population to avoid bias.

sample size: The sample number in the subset of a population.

sampling distribution: The distribution obtained by computing a statistic for all possible samples of the selected size drawn from the same population.

sampling variability: The tendency of the same statistic computed from a number of random samples drawn from the same population to differ.

scatter plot: A graphic display used to illustrate degree of correlation between two variables.

skewed: A distribution displaced at one end of the scale and a tail strung out at the other end.

slope: A measure of a line's slant.

spread: Identifies how similar or varied the set of values are for variables in a sample or population.

standard deviation: A measure of data variation; the square root of the variance.

standard error: A measure of the random variability of a statistic, such as the mean (i.e., standard error of the mean). The standard error of the mean is equal to the standard deviation divided by the square root of the sample size (n).

standardize: To convert to a z-score.

statistic: A characteristic of a sample. A statistic is an estimate of a population parameter. For larger samples, the statistic is a better estimate of the parameter.

statistical significance: The probability of obtaining a given result by chance. High statistical significance does not necessarily imply importance.

statistics: A branch of mathematics that describes and reasons from numerical observations or descriptive measures of a sample.

stem-and-leaf plot: A graphic display similar to a histogram that displays the shape of the data according to the first digit (stem) and the last digits (leaf). In a stem-and-leaf plot of test scores, for example, a score of 69 is depicted as follows: 6 is the stem and 9 is the leaf; for scores of 75, 77, and 79, 7 is the stem, and 5, 7, and 9 are the leaves.

symmetric distribution (perfect symmetry): A probability or frequency distribution that has the property in which the mean, median, and mode are all the same value.

symmetry: A shape such that one side is the exact mirror image of the other.

systematic error: The consistent underestimation or overestimation of a true value due to poor sampling technique. See also *bias*.

***t*-distribution:** A probability distribution often used when the population standard deviation is not known or when the sample size is small.

***t*-test:** A t-test is widely used in behavioral sciences to determine whether there is a statistically significant difference in the mean of a population from the mean of another population. It deals with inferences based on small samples (<30) when the population standard deviation is unknown.

tabled value: The value of a computed statistic used as a threshold to decide whether the null hypothesis will be rejected. Also called *critical value*.

test statistic: A computed quantity used to decide hypothesis tests.

two-tailed test: A test of the prediction that two values are equal, or a test that they are not equal. Also called a *nondirectional test*.

Type I error: Rejecting a null hypothesis that is, in fact, true. Represented by the Greek letter alpha, α.

Type II error: Failing to reject a null hypothesis that is, in fact, false. Represented by the Greek letter beta, β.

upper quartile: (Q_3), the 75th percentile of a set of measures.

value: A measurement or classification of a variable.

variable: An observable characteristic of a phenomenon that can be measured or classified.

variance: The sum of squared deviations of (n) measurements from their mean divided by ($n - 1$).

***z*-score:** A unit of measurement obtained by subtracting the mean from the value of the element and dividing by the standard deviation. It describes how many standard deviations the score is above or below the mean.

***z*-test:** A method of testing the null hypothesis whereby the test statistics are normally distributed given that the null hypothesis is true.

Appendix A
COMMON MISTAKES

It may seem that there are many ways to make errors in working a statistics problem. But in fact, most errors on statistics exams can be reduced to a short list of common oversights. If you learn to avoid the mistakes listed here, you can greatly reduce your chances of making an error on an exam.

- **Forgetting to convert between standard deviation (σ and s) and variance (σ^2 and s^2):** Some formulas use one; some use the other. Square the standard deviation to get the variance, or take the positive square root of the variance to get the standard deviation.
- **Misstating one-tailed and two-tailed hypotheses:** If the hypothesis predicts simply that one value will be higher than another, it requires a one-tailed test. If, however, it predicts that two values will be different—that is, one value will be either higher *or* lower than another or that they will be equal—then use a two-tailed test. Make sure your null and alternative hypotheses together cover all possibilities—greater than, less than, and equal to.
- **Failing to split the alpha level for two-tailed tests:** If the overall significance level for the test is 0.05, then you must look up the critical (tabled) value for a probability of 0.025. The alpha level is always split when computing confidence intervals.
- **Misreading the standard normal (z) table:** As explained in Chapter 5, all standard normal tables do not have the same format, and it is important to know what area of the curve (or probability) the table presents as corresponding to a given z-score. Table B-2 in Appendix B gives the area of the curve lying at or below z. The area to the right of z (or the probability of obtaining a value above z) is simply 1 minus the tabled probability.
- **Using n instead of $n - 1$ degrees of freedom in one-sample t-tests:** Remember that you must subtract 1 from n in order to get the degrees-of-freedom parameter that you need in order to look up a value in the t-table.

- **Confusing confidence level with confidence interval:** The confidence level is the significance level of the test or the likelihood of obtaining a given result by chance. The confidence interval is a range of values between the lowest and highest values that the estimated parameter could take at a given confidence level.
- **Confusing interval width with margin of error:** A confidence interval is always a point estimate plus or minus a margin of error. The interval width is double that margin of error. If, for example, a population parameter is estimated to be 46 percent plus or minus 4 percent, the interval width is 8 percent ($50 - 42 = 8$).
- **Confusing statistics with parameters:** Parameters are characteristics of the population that you usually do not know; they are designated with Greek symbols (μ and σ). Statistics are characteristics of samples that you are usually able to compute. Although statistics correspond to parameters ($\overline{x}$ is the mean of a sample, as μ is the mean of a population), the two are not interchangeable; hence, you need to be careful and know which variables are parameters and which are statistics. You compute statistics in order to estimate parameters.
- **Confusing the addition rule with the multiplication rule:** When determining probability, the multiplication rule applies if all favorable outcomes must occur in a series of events. The addition rule applies when *at least one* success must occur in a series of events.
- **Confusing the correlation relationship with the causation relationship:** Causation describes a chain of events that occurs between two things so that the first one *causes* the second one. Causation involves time. Correlation is a technique that can show only how strongly pairs of variables are related.
- **Forgetting that the addition rule applies to mutually exclusive outcomes:** If outcomes can occur together, the probability of their joint occurrence must be subtracted from the total "addition rule" probability. (See Chapter 4 for more information.)
- **Forgetting to average the two middle values of an even-numbered set when figuring median and (sometimes) quartiles:** If an ordered series contains an even number of measures, the median is always the mean of the two middle measures.
- **Forgetting where to place the points of a frequency polygon:** The points of a frequency polygon are always at the center of each of the class intervals, not at the ends.

Appendix B
FORMULAS AND TABLES

This appendix includes commonly used symbols and formulas and the following distribution tables:

Table B-1: Binomial Probabilities, *P(x)* for $n \leq 20$

Table B-2: Standard Normal Probabilities

Table B-3: *t*-distribution Critical Values

Table B-4: The Chi-Square Distribution: χ^2 Critical Values

Notation Symbols

Sample	***Population***
n = sample size	N = *population size*
$\overline{x}$ = sample mean	μ = *population mean*
s = sample standard deviation	σ = population standard deviation
s^2 = sample variance	σ^2 = population variance
p = *sample* proportion	π = population proportion
Δ = difference between the sample means	
Other Notation Conceptual Symbols	
Σ = sum of scores	df = degrees of freedom
$\cup$ = union (reads "or")	$\cap$ = intersection (reads "and")
α = type I error	β = type II error
H_0 = null hypothesis	H_a = alternate hypothesis

Formulas

Descriptive statistics

Sample	***Population***
Mean $\overline{x} = \dfrac{\Sigma x}{n}$	Mean $\mu = \dfrac{\Sigma x}{n}$
Mean (grouped data in frequency table) $\overline{x} = \dfrac{\Sigma(f \cdot x)}{\Sigma f}$	
Standard deviation $s = \sqrt{\dfrac{\Sigma(x-\overline{x})^2}{n-1}}$	Standard deviation $\sigma = \sqrt{\dfrac{\Sigma(x-\mu)^2}{N}}$ or $\sigma = \sqrt{\dfrac{\Sigma x^2}{N} - \mu^2}$
Standard deviation (grouped data in frequency table) $s = \sqrt{\dfrac{n\left[\Sigma(f \cdot x^2)\right]-\left[\Sigma(f \cdot x)\right]^2}{n(n-1)}}$	
Variance $s^2 = \dfrac{\Sigma(\overline{X}-X_1)^2}{N}$	Variance $\sigma^2 = \dfrac{\Sigma(X-\mu)^2}{N}$

Probability

If A, B are mutually exclusive	$P(A \cup B) = P(A) + P(B)$
If A, B are not mutually exclusive	$P(A \cup B) = P(A) + P(B) - P(A \cap B)$
If A, B are independent	$P(A \cap B) = P(A) \times P(B)$
If A, B are dependent	$P(A \cap B) = P(A) \times P(B \mid A)$
Rule of complements	$P(A) = 1 - P(B)$
Bayes' theorem	$P(A \mid B) = \dfrac{P(B \mid A) \cdot P(A)}{P(B)}$

Probability distributions

Mean (probability distribution)	$\mu = \Sigma[x \cdot P(x)]$
Standard deviation (probability distribution)	$\sigma = \sqrt{\Sigma\left[x^2 \cdot P(x)\right] - \mu^2}$
Binomial probability	$P(x) = \frac{n!}{(n-x)!x!} \cdot \pi^x \cdot 1 - \pi^{n-x}$
Mean (binomial)	$\mu = n \cdot \pi$
Variance (binomial)	$\sigma^2 = n \cdot \pi \cdot (\pi - 1)$
Standard deviation (binomial)	$\sigma = \sqrt{n \cdot \pi \cdot (\pi - 1)}$

Normal distribution

Standard score	$z = \frac{x - \mu}{\sigma}$ or $\frac{x - \overline{x}}{s}$
Central limit theorem	$\mu_{\overline{x}} = \mu$
Central limit theorem (standard error)	$\sigma_{\overline{x}} = \frac{\sigma}{\sqrt{n}}$

Confidence intervals (one population)

Confidence intervals for comparing two means	$(a,b) = \overline{x}_1 - \overline{x}_2 \pm t_{a/2} \cdot \sqrt{\frac{s_1^2}{n_1} \cdot \frac{s_2^2}{n_2}}$
Pooled variance	$s_p^2 = \frac{(n_1 - 1)s_1^2 + (n_2 - 1)s_2^2}{n_1 + n_2 - 2}$

Test statistics

Mean (one population unknown)	$t=\dfrac{\overline{x}-\mu}{\dfrac{s}{\sqrt{n}}}$
Mean (one population known)	$z=\dfrac{\overline{x}-\mu}{\dfrac{\sigma}{\sqrt{n}}}$
Variance	$X^2=\dfrac{(n-1)s^2}{\sigma^2}$
z-test (for two means)	$z=\dfrac{\overline{x}_1-\overline{x}_2-\mu}{\sqrt{\dfrac{\sigma_1^2}{n_1}+\dfrac{\sigma_2^2}{n_2}}}$
t-test (for two means)	$t=\dfrac{\overline{x}_1-\overline{x}_2-\mu}{\sqrt{\dfrac{s_1^2}{n_1}+\dfrac{s_2^2}{n_2}}}$

Pearson correlation

Conceptual equation for Pearson *r*	$r=\dfrac{\sum xy-\frac{1}{n}\left(\sum x\right)\left(\sum y\right)}{(n-1)s_x s_y}$
Pearson's product coefficient (*r*)	$r=\dfrac{\sum\left(\dfrac{x_i-\overline{x}}{s_x}\right)\left(\dfrac{y_i-\overline{y}}{s_y}\right)}{n-1}$
Significance of (*r*)	$t=\dfrac{r\sqrt{(n-2)}}{\sqrt{(1-r^2)}}$
Confidence interval for the slope	$(a,b)=\hat{\beta}_1\pm t_{a/2,n-2}\cdot\dfrac{s}{\sqrt{S_{xx}}}$
Confidence interval for prediction	$(a,b)=\hat{y}\pm t_{a/2,n-1}\left(\sqrt{1+\dfrac{1}{n}+\dfrac{(x-\overline{x})^2}{S_{xx}}}\right)$

Table B-1: Binomial Probabilities, $P(x)$ for $n \leq 20$

**n = number of events*

n = 2

					π						
x↓	***0.05***	***0.10***	***0.15***	***0.20***	***0.25***	***0.30***	***0.35***	***0.40***	***0.45***	***0.50***	
0	0.9025	0.8100	0.7225	0.6400	0.5625	0.4900	0.4225	0.3600	0.3025	0.2500	2
1	0.0950	0.1800	0.2550	0.3200	0.3750	0.4200	0.4550	0.4800	0.4950	0.5000	1
2	0.0025	0.0100	0.0225	0.0400	0.0625	0.0900	0.1225	0.1600	0.2025	0.2500	0
	0.95	0.90	0.85	0.80	0.75	0.70	0.65	0.60	0.55	0.50	*x*↑

n = 3

					π						
x↓	***0.05***	***0.10***	***0.15***	***0.20***	***0.25***	***0.30***	***0.35***	***0.40***	***0.45***	***0.50***	
0	0.8574	0.7290	0.6141	0.5120	0.4219	0.3430	0.2746	0.2160	0.1664	0.1250	3
1	0.1354	0.2430	0.3251	0.3840	0.4219	0.4410	0.4436	0.4320	0.4084	0.3750	2
2	0.0071	0.0270	0.0574	0.0960	0.1406	0.1890	0.2389	0.2880	0.3341	0.3750	1
3	0.0001	0.0010	0.0034	0.0080	0.0156	0.0270	0.0429	0.0640	0.0911	0.1250	0
	0.95	0.90	0.85	0.80	0.75	0.70	0.65	0.60	0.55	0.50	*x*↑

***n* = 4**

					π						
x↓	**0.05**	**0.10**	**0.15**	**0.20**	**0.25**	**0.30**	**0.35**	**0.40**	**0.45**	**0.50**	
0	0.8145	0.6561	0.5220	0.4096	0.3164	0.2401	0.1785	0.1296	0.0915	0.0625	4
1	0.1715	0.2916	0.3685	0.4096	0.4219	0.4116	0.3845	0.3456	0.2995	0.2500	3
2	0.0135	0.0486	0.0975	0.1536	0.2109	0.2646	0.3105	0.3456	0.3675	0.3750	2
3	0.0005	0.0036	0.0115	0.0256	0.0469	0.0756	0.1115	0.1536	0.2005	0.2500	1
4	0.0000	0.0001	0.0005	0.0016	0.0039	0.0081	0.0150	0.0256	0.0410	0.0625	0
	0.95	0.90	0.85	0.80	0.75	0.70	0.65	0.60	0.55	0.50	*x*↑

***n* = 5**

					π						
x↓	**0.05**	**0.10**	**0.15**	**0.20**	**0.25**	**0.30**	**0.35**	**0.40**	**0.45**	**0.50**	
0	0.7738	0.5905	0.4437	0.3277	0.2373	0.1681	0.1160	0.0778	0.0503	0.0313	5
1	0.2036	0.3281	0.3915	0.4096	0.3955	0.3602	0.3124	0.2592	0.2059	0.1563	4
2	0.0214	0.0729	0.1382	0.2048	0.2637	0.3087	0.3364	0.3456	0.3369	0.3125	3
3	0.0011	0.0081	0.0244	0.0512	0.0879	0.1323	0.1811	0.2304	0.2757	0.3125	2
4	0.0000	0.0005	0.0022	0.0064	0.0146	0.0284	0.0488	0.0768	0.1128	0.1563	1
5	0.0000	0.0000	0.0001	0.0003	0.0010	0.0024	0.0053	0.0102	0.0185	0.0313	0
	0.95	0.90	0.85	0.80	0.75	0.70	0.65	0.60	0.55	0.50	x↑

$n = 6$											
					π						
x↓	***0.05***	***0.10***	***0.15***	***0.20***	***0.25***	***0.30***	***0.35***	***0.40***	***0.45***	***0.50***	
0	0.7351	0.5314	0.3771	0.2621	0.1780	0.1176	0.0754	0.0467	0.0277	0.0156	6
1	0.2321	0.3543	0.3993	0.3932	0.3560	0.3025	0.2437	0.1866	0.1359	0.0938	5
2	0.0305	0.0984	0.1762	0.2458	0.2966	0.3241	0.3280	0.3110	0.2780	0.2344	4
3	0.0021	0.0146	0.0415	0.0819	0.1318	0.1852	0.2355	0.2765	0.3032	0.3125	3
4	0.0001	0.0012	0.0055	0.0154	0.0330	0.0595	0.0951	0.1382	0.1861	0.2344	2
5	0.0000	0.0001	0.0004	0.0015	0.0044	0.0102	0.0205	0.0369	0.0609	0.0938	1
6	0.0000	0.0000	0.0000	0.0001	0.0002	0.0007	0.0018	0.0041	0.0083	0.0156	0
	0.95	0.90	0.85	0.80	0.75	0.70	0.65	0.60	0.55	0.50	*x*↑

n = 7

	π										
x↓	**0.05**	**0.10**	**0.15**	**0.20**	**0.25**	**0.30**	**0.35**	**0.40**	**0.45**	**0.50**	
0	0.6983	0.4783	0.3206	0.2097	0.1335	0.0824	0.0490	0.0280	0.0152	0.0078	7
1	0.2573	0.3720	0.3960	0.3670	0.3115	0.2471	0.1848	0.1306	0.0872	0.0547	6
2	0.0406	0.1240	0.2097	0.2753	0.3115	0.3177	0.2985	0.2613	0.2140	0.1641	5
3	0.0036	0.0230	0.0617	0.1147	0.1730	0.2269	0.2679	0.2903	0.2918	0.2734	4
4	0.0002	0.0026	0.0109	0.0287	0.0577	0.0972	0.1442	0.1935	0.2388	0.2734	3
5	0.0000	0.0002	0.0012	0.0043	0.0115	0.0250	0.0466	0.0774	0.1172	0.1641	2
6	0.0000	0.0000	0.0001	0.0004	0.0013	0.0036	0.0084	0.0172	0.0320	0.0547	1
7	0.0000	0.0000	0.0000	0.0000	0.0001	0.0002	0.0006	0.0016	0.0037	0.0078	0
	0.95	0.90	0.85	0.80	0.75	0.70	0.65	0.60	0.55	0.50	x↑

***n* = 8**

					π						
x↓	**0.05**	**0.10**	**0.15**	**0.20**	**0.25**	**0.30**	**0.35**	**0.40**	**0.45**	**0.50**	
0	0.6634	0.4305	0.2725	0.1678	0.1001	0.0576	0.0319	0.0168	0.0084	0.0039	8
1	0.2793	0.3826	0.3847	0.3355	0.2670	0.1977	0.1373	0.0896	0.0548	0.0313	7
2	0.0515	0.1488	0.2376	0.2936	0.3115	0.2965	0.2587	0.2090	0.1569	0.1094	6
3	0.0054	0.0331	0.0839	0.1468	0.2076	0.2541	0.2786	0.2787	0.2568	0.2188	5
4	0.0004	0.0046	0.0185	0.0459	0.0865	0.1361	0.1875	0.2322	0.2627	0.2734	4
5	0.0000	0.0004	0.0026	0.0092	0.0231	0.0467	0.0808	0.1239	0.1719	0.2188	3
6	0.0000	0.0000	0.0002	0.0011	0.0038	0.0100	0.0217	0.0413	0.0703	0.1094	2
7	0.0000	0.0000	0.0000	0.0001	0.0004	0.0012	0.0033	0.0079	0.0164	0.0313	1
8	0.0000	0.0000	0.0000	0.0000	0.0000	0.0001	0.0002	0.0007	0.0017	0.0039	0
	0.95	0.90	0.85	0.80	0.75	0.70	0.65	0.60	0.55	0.50	x↑

***n* = 9**

					π						
x↓	***0.05***	***0.10***	***0.15***	***0.20***	***0.25***	***0.30***	***0.35***	***0.40***	***0.45***	***0.50***	
0	0.6302	0.3874	0.2316	0.1342	0.0751	0.0404	0.0207	0.0101	0.0046	0.0020	9
1	0.2985	0.3874	0.3679	0.3020	0.2253	0.1556	0.1004	0.0605	0.0339	0.0176	8
2	0.0629	0.1722	0.2597	0.3020	0.3003	0.2668	0.2162	0.1612	0.1110	0.0703	7
3	0.0077	0.0446	0.1069	0.1762	0.2336	0.2668	0.2716	0.2508	0.2119	0.1641	6
4	0.0006	0.0074	0.0283	0.0661	0.1168	0.1715	0.2194	0.2508	0.2600	0.2461	5
5	0.0000	0.0008	0.0050	0.0165	0.0389	0.0735	0.1181	0.1672	0.2128	0.2461	4
6	0.0000	0.0001	0.0006	0.0028	0.0087	0.0210	0.0424	0.0743	0.1160	0.1641	3
7	0.0000	0.0000	0.0000	0.0003	0.0012	0.0039	0.0098	0.0212	0.0407	0.0703	2
8	0.0000	0.0000	0.0000	0.0000	0.0001	0.0004	0.0013	0.0035	0.0083	0.0176	1
9	0.0000	0.0000	0.0000	0.0000	0.0000	0.0000	0.0001	0.0003	0.0008	0.0020	0
	0.95	0.90	0.85	0.80	0.75	0.70	0.65	0.60	0.55	0.50	*x*↑

***n* = 10**

					π						
x↓	**0.05**	**0.10**	**0.15**	**0.20**	**0.25**	**0.30**	**0.35**	**0.40**	**0.45**	**0.50**	
0	0.5987	0.3487	0.1969	0.1074	0.0563	0.0282	0.0135	0.0060	0.0025	0.0010	10
1	0.3151	0.3874	0.3474	0.2684	0.1877	0.1211	0.0725	0.0403	0.0207	0.0098	9
2	0.0746	0.1937	0.2759	0.3020	0.2816	0.2335	0.1757	0.1209	0.0763	0.0439	8
3	0.0105	0.0574	0.1298	0.2013	0.2503	0.2668	0.2522	0.2150	0.1665	0.1172	7
4	0.0010	0.0112	0.0401	0.0881	0.1460	0.2001	0.2377	0.2508	0.2384	0.2051	6
5	0.0001	0.0015	0.0085	0.0264	0.0584	0.1029	0.1536	0.2007	0.2340	0.2461	5
6	0.0000	0.0001	0.0012	0.0055	0.0162	0.0368	0.0689	0.1115	0.1596	0.2051	4
7	0.0000	0.0000	0.0001	0.0008	0.0031	0.0090	0.0212	0.0425	0.0746	0.1172	3
8	0.0000	0.0000	0.0000	0.0001	0.0004	0.0014	0.0043	0.0106	0.0229	0.0439	2
9	0.0000	0.0000	0.0000	0.0000	0.0000	0.0001	0.0005	0.0016	0.0042	0.0098	1
10	0.0000	0.0000	0.0000	0.0000	0.0000	0.0000	0.0000	0.0001	0.0003	0.0010	0
	0.95	0.90	0.85	0.80	0.75	0.70	0.65	0.60	0.55	0.50	x↑

***n* = 12**

					π						
x↓	**0.05**	**0.10**	**0.15**	**0.20**	**0.25**	**0.30**	**0.35**	**0.40**	**0.45**	**0.50**	
0	0.5404	0.2824	0.1422	0.0687	0.0317	0.0138	0.0057	0.0022	0.0008	0.0002	12
1	0.3413	0.3766	0.3012	0.2062	0.1267	0.0712	0.0368	0.0174	0.0075	0.0029	11
2	0.0988	0.2301	0.2924	0.2835	0.2323	0.1678	0.1088	0.0639	0.0339	0.0161	10
3	0.0173	0.0852	0.1720	0.2362	0.2581	0.2397	0.1954	0.1419	0.0923	0.0537	9
4	0.0021	0.0213	0.0683	0.1329	0.1936	0.2311	0.2367	0.2128	0.1700	0.1208	8
5	0.0002	0.0038	0.0193	0.0532	0.1032	0.1585	0.2039	0.2270	0.2225	0.1934	7
6	0.0000	0.0005	0.0040	0.0155	0.0401	0.0792	0.1281	0.1766	0.2124	0.2256	6
7	0.0000	0.0000	0.0006	0.0033	0.0115	0.0291	0.0591	0.1009	0.1489	0.1934	5
8	0.0000	0.0000	0.0001	0.0005	0.0024	0.0078	0.0199	0.0420	0.0762	0.1208	4
9	0.0000	0.0000	0.0000	0.0001	0.0004	0.0015	0.0048	0.0125	0.0277	0.0537	3
10	0.0000	0.0000	0.0000	0.0000	0.0000	0.0002	0.0008	0.0025	0.0068	0.0161	2
11	0.0000	0.0000	0.0000	0.0000	0.0000	0.0000	0.0001	0.0003	0.0010	0.0029	1
12	0.0000	0.0000	0.0000	0.0000	0.0000	0.0000	0.0000	0.0000	0.0001	0.0002	0
	0.95	0.90	0.85	0.80	0.75	0.70	0.65	0.60	0.55	0.50	x↑

***n* = 14**

					π						
x↓	**0.05**	**0.10**	**0.15**	**0.20**	**0.25**	**0.30**	**0.35**	**0.40**	**0.45**	**0.50**	
0	0.4877	0.2288	0.1028	0.0440	0.0178	0.0068	0.0024	0.0008	0.0002	0.0001	14
1	0.3593	0.3559	0.2539	0.1539	0.0832	0.0407	0.0181	0.0073	0.0027	0.0009	13
2	0.1229	0.2570	0.2912	0.2501	0.1802	0.1134	0.0634	0.0317	0.0141	0.0056	12
3	0.0259	0.1142	0.2056	0.2501	0.2402	0.1943	0.1366	0.0845	0.0462	0.0222	11
4	0.0037	0.0349	0.0998	0.1720	0.2202	0.2290	0.2022	0.1549	0.1040	0.0611	10
5	0.0004	0.0078	0.0352	0.0860	0.1468	0.1963	0.2178	0.2066	0.1701	0.1222	9
6	0.0000	0.0013	0.0093	0.0322	0.0734	0.1262	0.1759	0.2066	0.2088	0.1833	8
7	0.0000	0.0002	0.0019	0.0092	0.0280	0.0618	0.1082	0.1574	0.1952	0.2095	7
8	0.0000	0.0000	0.0003	0.0020	0.0082	0.0232	0.0510	0.0918	0.1398	0.1833	6
9	0.0000	0.0000	0.0000	0.0003	0.0018	0.0066	0.0183	0.0408	0.0762	0.1222	5
10	0.0000	0.0000	0.0000	0.0000	0.0003	0.0014	0.0049	0.0136	0.0312	0.0611	4
11	0.0000	0.0000	0.0000	0.0000	0.0000	0.0002	0.0010	0.0033	0.0093	0.0222	3
12	0.0000	0.0000	0.0000	0.0000	0.0000	0.0000	0.0001	0.0005	0.0019	0.0056	2
13	0.0000	0.0000	0.0000	0.0000	0.0000	0.0000	0.0000	0.0001	0.0002	0.0009	1
14	0.0000	0.0000	0.0000	0.0000	0.0000	0.0000	0.0000	0.0000	0.0000	0.0001	0
	0.95	0.90	0.85	0.80	0.75	0.70	0.65	0.60	0.55	0.50	x↑

***n* = 16**

	π										
x↓	***0.05***	***0.10***	***0.15***	***0.20***	***0.25***	***0.30***	***0.35***	***0.40***	***0.45***	***0.50***	
0	0.4401	0.1853	0.0743	0.0281	0.0100	0.0033	0.0010	0.0003	0.0001	0.0000	16
1	0.3706	0.3294	0.2097	0.1126	0.0535	0.0228	0.0087	0.0030	0.0009	0.0002	15
2	0.1463	0.2745	0.2775	0.2111	0.1336	0.0732	0.0353	0.0150	0.0056	0.0018	14
3	0.0359	0.1423	0.2285	0.2463	0.2079	0.1465	0.0888	0.0468	0.0215	0.0085	13
4	0.0061	0.0514	0.1311	0.2001	0.2252	0.2040	0.1553	0.1014	0.0572	0.0278	12
5	0.0008	0.0137	0.0555	0.1201	0.1802	0.2099	0.2008	0.1623	0.1123	0.0667	11
6	0.0001	0.0028	0.0180	0.0550	0.1101	0.1649	0.1982	0.1983	0.1684	0.1222	10
7	0.0000	0.0004	0.0045	0.0197	0.0524	0.1010	0.1524	0.1889	0.1969	0.1746	9
8	0.0000	0.0001	0.0009	0.0055	0.0197	0.0487	0.0923	0.1417	0.1812	0.1964	8
9	0.0000	0.0000	0.0001	0.0012	0.0058	0.0185	0.0442	0.0840	0.1318	0.1746	7
10	0.0000	0.0000	0.0000	0.0002	0.0014	0.0056	0.0167	0.0392	0.0755	0.1222	6
11	0.0000	0.0000	0.0000	0.0000	0.0002	0.0013	0.0049	0.0142	0.0337	0.0667	5
12	0.0000	0.0000	0.0000	0.0000	0.0000	0.0002	0.0011	0.0040	0.0115	0.0278	4
13	0.0000	0.0000	0.0000	0.0000	0.0000	0.0000	0.0002	0.0008	0.0029	0.0085	3
14	0.0000	0.0000	0.0000	0.0000	0.0000	0.0000	0.0000	0.0001	0.0005	0.0018	2
15	0.0000	0.0000	0.0000	0.0000	0.0000	0.0000	0.0000	0.0000	0.0001	0.0002	1
	0.95	0.90	0.85	0.80	0.75	0.70	0.65	0.60	0.55	0.50	*x*↑

***n* = 18**

					π						
x↓	***0.05***	***0.10***	***0.15***	***0.20***	***0.25***	***0.30***	***0.35***	***0.40***	***0.45***	***0.50***	
0	0.3972	0.1501	0.0536	0.0180	0.0056	0.0016	0.0004	0.0001	0.0000	0.0000	18
1	0.3763	0.3002	0.1704	0.0811	0.0338	0.0126	0.0042	0.0012	0.0003	0.0001	17
2	0.1683	0.2835	0.2556	0.1723	0.0958	0.0458	0.0190	0.0069	0.0022	0.0006	16
3	0.0473	0.1680	0.2406	0.2297	0.1704	0.1046	0.0547	0.0246	0.0095	0.0031	15
4	0.0093	0.0700	0.1592	0.2153	0.2130	0.1681	0.1104	0.0614	0.0291	0.0117	14
5	0.0014	0.0218	0.0787	0.1507	0.1988	0.2017	0.1664	0.1146	0.0666	0.0327	13
6	0.0002	0.0052	0.0301	0.0816	0.1436	0.1873	0.1941	0.1655	0.1181	0.0708	12
7	0.0000	0.0010	0.0091	0.0350	0.0820	0.1376	0.1792	0.1892	0.1657	0.1214	11
8	0.0000	0.0002	0.0022	0.0120	0.0376	0.0811	0.1327	0.1734	0.1864	0.1669	10
9	0.0000	0.0000	0.0004	0.0033	0.0139	0.0386	0.0794	0.1284	0.1694	0.1855	9
10	0.0000	0.0000	0.0001	0.0008	0.0042	0.0149	0.0385	0.0771	0.1248	0.1669	8
11	0.0000	0.0000	0.0000	0.0001	0.0010	0.0046	0.0151	0.0374	0.0742	0.1214	7
12	0.0000	0.0000	0.0000	0.0000	0.0002	0.0012	0.0047	0.0145	0.0354	0.0708	6
13	0.0000	0.0000	0.0000	0.0000	0.0000	0.0002	0.0012	0.0045	0.0134	0.0327	5
14	0.0000	0.0000	0.0000	0.0000	0.0000	0.0000	0.0002	0.0011	0.0039	0.0117	4
15	0.0000	0.0000	0.0000	0.0000	0.0000	0.0000	0.0000	0.0002	0.0009	0.0031	3
16	0.0000	0.0000	0.0000	0.0000	0.0000	0.0000	0.0000	0.0000	0.0001	0.0006	2
17	0.0000	0.0000	0.0000	0.0000	0.0000	0.0000	0.0000	0.0000	0.0000	0.0001	1
	0.95	0.90	0.85	0.80	0.75	0.70	0.65	0.60	0.55	0.50	*x*↑

***n* = 20**

					π						
x↓	**0.05**	**0.10**	**0.15**	**0.20**	**0.25**	**0.30**	**0.35**	**0.40**	**0.45**	**0.50**	
0	0.3585	0.1216	0.0388	0.0115	0.0032	0.0008	0.0002	0.0000	0.0000	0.0000	20
1	0.3774	0.2702	0.1368	0.0576	0.0211	0.0068	0.0020	0.0005	0.0001	0.0000	19
2	0.1887	0.2852	0.2293	0.1369	0.0669	0.0278	0.0100	0.0031	0.0008	0.0002	18
3	0.0596	0.1901	0.2428	0.2054	0.1339	0.0716	0.0323	0.0123	0.0040	0.0011	17
4	0.0133	0.0898	0.1821	0.2182	0.1897	0.1304	0.0738	0.0350	0.0139	0.0046	16
5	0.0022	0.0319	0.1028	0.1746	0.2023	0.1789	0.1272	0.0746	0.0365	0.0148	15
6	0.0003	0.0089	0.0454	0.1091	0.1686	0.1916	0.1712	0.1244	0.0746	0.0370	14
7	0.0000	0.0020	0.0160	0.0545	0.1124	0.1643	0.1844	0.1659	0.1221	0.0739	13
8	0.0000	0.0004	0.0046	0.0222	0.0609	0.1144	0.1614	0.1797	0.1623	0.1201	12
9	0.0000	0.0001	0.0011	0.0074	0.0271	0.0654	0.1158	0.1597	0.1771	0.1602	11
10	0.0000	0.0000	0.0002	0.0020	0.0099	0.0308	0.0686	0.1171	0.1593	0.1762	10
11	0.0000	0.0000	0.0000	0.0005	0.0030	0.0120	0.0336	0.0710	0.1185	0.1602	9
12	0.0000	0.0000	0.0000	0.0001	0.0008	0.0039	0.0136	0.0355	0.0727	0.1201	8
13	0.0000	0.0000	0.0000	0.0000	0.0002	0.0010	0.0045	0.0146	0.0366	0.0739	7
14	0.0000	0.0000	0.0000	0.0000	0.0000	0.0002	0.0012	0.0049	0.0150	0.0370	6
15	0.0000	0.0000	0.0000	0.0000	0.0000	0.0000	0.0003	0.0013	0.0049	0.0148	5
16	0.0000	0.0000	0.0000	0.0000	0.0000	0.0000	0.0000	0.0003	0.0013	0.0046	4
17	0.0000	0.0000	0.0000	0.0000	0.0000	0.0000	0.0000	0.0000	0.0002	0.0011	3
18	0.0000	0.0000	0.0000	0.0000	0.0000	0.0000	0.0000	0.0000	0.0000	0.0002	2
	0.95	0.90	0.85	0.80	0.75	0.70	0.65	0.60	0.55	0.50	x↑

Table B-2: Standard Normal Probabilities

(Table entry is probability at or below *z*.)

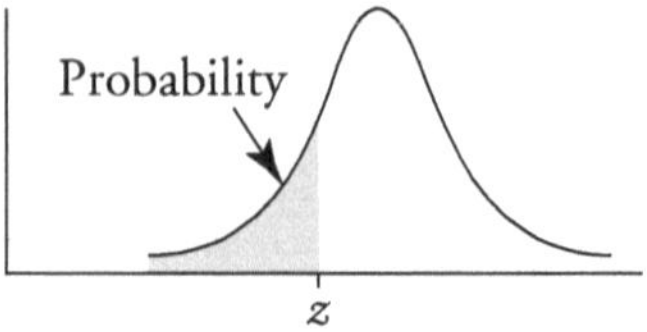

z	0.00	0.01	0.02	0.03	0.04
−3.4	0.0003	0.0003	0.0003	0.0003	0.0003
−3.3	0.0005	0.0005	0.0005	0.0004	0.0004
−3.2	0.0007	0.0007	0.0006	0.0006	0.0006
−3.1	0.0010	0.0009	0.0009	0.0009	0.0008
−3.0	0.0013	0.0013	0.0013	0.0012	0.0012
−2.9	0.0019	0.0018	0.0018	0.0017	0.0016
−2.8	0.0026	0.0025	0.0024	0.0023	0.0023
−2.7	0.0035	0.0034	0.0033	0.0032	0.0031
−2.6	0.0047	0.0045	0.0044	0.0043	0.0041
−2.5	0.0062	0.0060	0.0059	0.0057	0.0055
−2.4	0.0082	0.0080	0.0078	0.0075	0.0073
−2.3	0.0107	0.0104	0.0102	0.0099	0.0096
−2.2	0.0139	0.0136	0.0132	0.0129	0.0125
−2.1	0.0179	0.0174	0.0170	0.0166	0.0162
−2.0	0.0228	0.0222	0.0217	0.0212	0.0207
−1.9	0.0287	0.0281	0.0274	0.0268	0.0262
−1.8	0.0359	0.0351	0.0344	0.0336	0.0329
−1.7	0.0446	0.0436	0.0427	0.0418	0.0409
−1.6	0.0548	0.0537	0.0526	0.0516	0.0505
−1.5	0.0668	0.0655	0.0643	0.0630	0.0618
−1.4	0.0808	0.0793	0.0778	0.0764	0.0749
−1.3	0.0968	0.0951	0.0934	0.0918	0.0901
−1.2	0.1151	0.1131	0.1112	0.1093	0.1075
−1.1	0.1357	0.1335	0.1314	0.1292	0.1271
−1.0	0.1587	0.1562	0.1539	0.1515	0.1492
−0.9	0.1841	0.1814	0.1788	0.1762	0.1736
−0.8	0.2119	0.2090	0.2061	0.2033	0.2005
−0.7	0.2420	0.2389	0.2358	0.2327	0.2296
−0.6	0.2743	0.2709	0.2676	0.2643	0.2611
−0.5	0.3085	0.3050	0.3015	0.2981	0.2946
−0.4	0.3446	0.3409	0.3372	0.3336	0.3300
−0.3	0.3821	0.3783	0.3745	0.3707	0.3669
−0.2	0.4207	0.4168	0.4129	0.4090	0.4052
−0.1	0.4602	0.4562	0.4522	0.4483	0.4443
−0.0	0.5000	0.4960	0.4920	0.4880	0.4840

0.05	0.06	0.07	0.08	0.09	z
0.0003	0.0003	0.0003	0.0003	0.0002	−3.4
0.0004	0.0004	0.0004	0.0004	0.0003	−3.3
0.0006	0.0006	0.0005	0.0005	0.0005	−3.2
0.0008	0.0008	0.0008	0.0007	0.0007	−3.1
0.0011	0.0011	0.0011	0.0010	0.0010	−3.0
0.0016	0.0015	0.0015	0.0014	0.0014	−2.9
0.0022	0.0021	0.0021	0.0020	0.0019	−2.8
0.0030	0.0029	0.0028	0.0027	0.0026	−2.7
0.0040	0.0039	0.0038	0.0037	0.0036	−2.6
0.0054	0.0052	0.0051	0.0049	0.0048	−2.5
0.0071	0.0069	0.0068	0.0066	0.0064	−2.4
0.0094	0.0091	0.0089	0.0087	0.0084	−2.3
0.0122	0.0119	0.0116	0.0113	0.0110	−2.2
0.0158	0.0154	0.0150	0.0146	0.0143	−2.1
0.0202	0.0197	0.0192	0.0188	0.0183	−2.0
0.0256	0.0250	0.0244	0.0239	0.0233	−1.9
0.0322	0.0314	0.0307	0.0301	0.0294	−1.8
0.0401	0.0392	0.0384	0.0375	0.0367	−1.7
0.0495	0.0485	0.0475	0.0465	0.0455	−1.6
0.0606	0.0594	0.0582	0.0571	0.0559	−1.5
0.0735	0.0721	0.0708	0.0694	0.0681	−1.4
0.0885	0.0869	0.0853	0.0838	0.0823	−1.3
0.1056	0.1038	0.1020	0.1003	0.0985	−1.2
0.1251	0.1230	0.1210	0.1190	0.1170	−1.1
0.1469	0.1446	0.1423	0.1401	0.1379	−1.0
0.1711	0.1685	0.1660	0.1635	0.1611	−0.9
0.1977	0.1949	0.1922	0.1894	0.1867	−0.8
0.2266	0.2236	0.2206	0.2177	0.2148	−0.7
0.2578	0.2546	0.2514	0.2483	0.2451	−0.6
0.2912	0.2877	0.2843	0.2810	0.2776	−0.5
0.3264	0.3228	0.3192	0.3156	0.3121	−0.4
0.3632	0.3594	0.3557	0.3520	0.3483	−0.3
0.4013	0.3974	0.3936	0.3897	0.3859	−0.2
0.4404	0.4364	0.4325	0.4286	0.4247	−0.1
0.4801	0.4761	0.4721	0.4681	0.4641	−0.0

Continued

Table B-2: Standard Normal Probabilities (Continued)

(Table entry is the probability at or below *z*.)

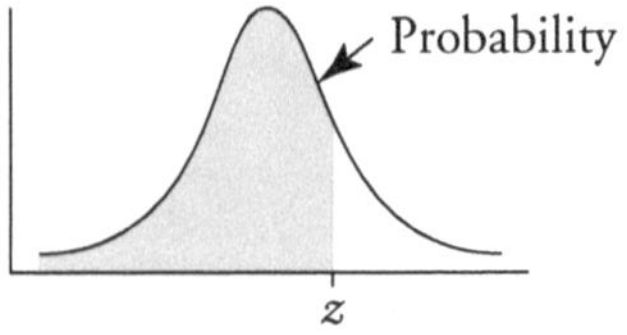

z	0.00	0.01	0.02	0.03	0.04
0.0	0.5000	0.5040	0.5080	0.5120	0.5160
0.1	0.5398	0.5438	0.5478	0.5517	0.5557
0.2	0.5793	0.5832	0.5871	0.5910	0.5948
0.3	0.6179	0.6217	0.6255	0.6293	0.6331
0.4	0.6554	0.6591	0.6628	0.6664	0.6700
0.5	0.6915	0.6950	0.6985	0.7019	0.7054
0.6	0.7257	0.7291	0.7324	0.7357	0.7389
0.7	0.7580	0.7611	0.7642	0.7673	0.7704
0.8	0.7881	0.7910	0.7939	0.7967	0.7995
0.9	0.8159	0.8186	0.8212	0.8238	0.8264
1.0	0.8413	0.8438	0.8461	0.8485	0.8508
1.1	0.8643	0.8665	0.8686	0.8708	0.8729
1.2	0.8849	0.8869	0.8888	0.8907	0.8925
1.3	0.9032	0.9049	0.9066	0.9082	0.9099
1.4	0.9192	0.9207	0.9222	0.9236	0.9251
1.5	0.9332	0.9345	0.9357	0.9370	0.9382
1.6	0.9452	0.9463	0.9474	0.9484	0.9495
1.7	0.9554	0.9564	0.9573	0.9582	0.9591
1.8	0.9641	0.9649	0.9656	0.9664	0.9671
1.9	0.9713	0.9719	0.9726	0.9732	0.9738
2.0	0.9772	0.9778	0.9783	0.9788	0.9793
2.1	0.9821	0.9826	0.9830	0.9834	0.9838
2.2	0.9861	0.9864	0.9868	0.9871	0.9875
2.3	0.9893	0.9896	0.9898	0.9901	0.9904
2.4	0.9918	0.9920	0.9922	0.9925	0.9927
2.5	0.9938	0.9940	0.9941	0.9943	0.9945
2.6	0.9953	0.9955	0.9956	0.9957	0.9959
2.7	0.9965	0.9966	0.9967	0.9968	0.9969
2.8	0.9974	0.9975	0.9976	0.9977	0.9977
2.9	0.9981	0.9982	0.9982	0.9983	0.9984
3.0	0.9987	0.9987	0.9987	0.9988	0.9988
3.1	0.9990	0.9991	0.9991	0.9991	0.9992
3.2	0.9993	0.9993	0.9994	0.9994	0.9994
3.3	0.9995	0.9995	0.9995	0.9996	0.9996
3.4	0.9997	0.9997	0.9997	0.9997	0.9997

0.05	*0.06*	*0.07*	*0.08*	*0.09*	*z*
0.5199	0.5239	0.5279	0.5319	0.5359	0.0
0.5596	0.5636	0.5675	0.5714	0.5753	0.1
0.5987	0.6026	0.6064	0.6103	0.6141	0.2
0.6368	0.6406	0.6443	0.6480	0.6517	0.3
0.6736	0.6772	0.6808	0.6844	0.6879	0.4
0.7088	0.7123	0.7157	0.7190	0.7224	0.5
0.7422	0.7454	0.7486	0.7517	0.7549	0.6
0.7734	0.7764	0.7794	0.7823	0.7852	0.7
0.8023	0.8051	0.8078	0.8106	0.8133	0.8
0.8289	0.8315	0.8340	0.8365	0.8389	0.9
0.8531	0.8554	0.8577	0.8599	0.8621	1.0
0.8749	0.8770	0.8790	0.8810	0.8830	1.1
0.8944	0.8962	0.8980	0.8997	0.9015	1.2
0.9115	0.9131	0.9147	0.9162	0.9177	1.3
0.9265	0.9279	0.9292	0.9306	0.9319	1.4
0.9394	0.9406	0.9418	0.9429	0.9441	1.5
0.9505	0.9515	0.9525	0.9535	0.9545	1.6
0.9599	0.9608	0.9616	0.9625	0.9633	1.7
0.9678	0.9686	0.9693	0.9699	0.9706	1.8
0.9744	0.9750	0.9756	0.9761	0.9767	1.9
0.9798	0.9803	0.9808	0.9812	0.9817	2.0
0.9842	0.9846	0.9850	0.9854	0.9857	2.1
0.9878	0.9881	0.9884	0.9887	0.9890	2.2
0.9906	0.9909	0.9911	0.9913	0.9916	2.3
0.9929	0.9931	0.9932	0.9934	0.9936	2.4
0.9946	0.9948	0.9949	0.9951	0.9952	2.5
0.9960	0.9961	0.9962	0.9963	0.9964	2.6
0.9970	0.9971	0.9972	0.9973	0.9974	2.7
0.9978	0.9979	0.9979	0.9980	0.9981	2.8
0.9984	0.9985	0.9985	0.9986	0.9986	2.9
0.9989	0.9989	0.9989	0.9990	0.9990	3.0
0.9992	0.9992	0.9992	0.9993	0.9993	3.1
0.9994	0.9994	0.9995	0.9995	0.9995	3.2
0.9996	0.9996	0.9996	0.9996	0.9997	3.3
0.9997	0.9997	0.9997	0.9997	0.9998	3.4

Table B-3: The *t*-distribution Critical Values

(Table entry is the value of point t corresponding to proportions in one tail with the given probability p lying above it.)

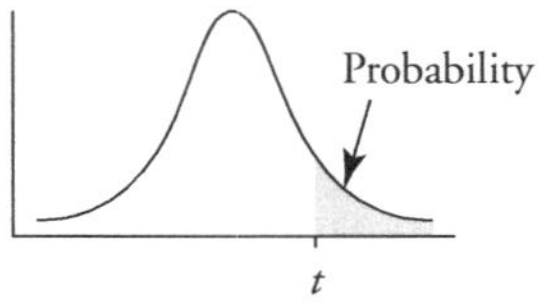

Note: One tail (either right or left)

Degrees of freedom (df)	Proportion in one tail											
	0.25	0.20	0.15	0.10	0.05	0.025	0.02	0.01	0.005	0.0025	0.001	0.0005
1	1.000	1.376	1.963	3.078	6.314	12.71	15.89	31.82	63.66	127.3	318.3	636.6
2	0.816	1.061	1.386	1.886	2.920	4.303	4.849	6.965	9.925	14.09	22.33	31.60
3	0.765	0.978	1.250	1.638	2.353	3.182	3.482	4.541	5.841	7.453	10.21	12.92
4	0.741	0.941	1.190	1.533	2.132	2.776	2.999	3.747	4.604	5.598	7.173	8.610
5	0.727	0.920	1.156	1.476	2.015	2.571	2.757	3.365	4.032	4.773	5.893	6.869
6	0.718	0.906	1.134	1.440	1.943	2.447	2.612	3.143	3.707	4.317	5.208	5.959
7	0.711	0.896	1.119	1.415	1.895	2.365	2.517	2.998	3.499	4.029	4.785	5.408
8	0.706	0.889	1.108	1.397	1.860	2.306	2.449	2.896	3.355	3.833	4.501	5.041
9	0.703	0.883	1.100	1.383	1.833	2.262	2.398	2.821	3.250	3.690	4.297	4.781
10	0.700	0.879	1.093	1.372	1.812	2.228	2.359	2.764	3.169	3.581	4.144	4.587
11	0.697	0.876	1.088	1.363	1.796	2.201	2.328	2.718	3.106	3.497	4.025	4.437
12	0.695	0.873	1.083	1.356	1.782	2.179	2.303	2.681	3.055	3.428	3.930	4.318
13	0.694	0.870	1.079	1.350	1.771	2.160	2.282	2.650	3.012	3.372	3.852	4.221
14	0.692	0.868	1.076	1.345	1.761	2.145	2.264	2.624	2.977	3.326	3.787	4.140

Degrees of freedom (df)	Proportion in one tail 0.25	0.20	0.15	0.10	0.05	0.025	0.02	0.01	0.005	0.0025	0.001	0.0005
15	0.691	0.866	1.074	1.341	1.753	2.131	2.249	2.602	2.947	3.286	3.733	4.073
16	0.690	0.865	1.071	1.337	1.746	2.120	2.235	2.583	2.921	3.252	3.686	4.015
17	0.689	0.863	1.069	1.333	1.740	2.110	2.224	2.567	2.898	3.222	3.646	3.965
18	0.688	0.862	1.067	1.330	1.734	2.101	2.214	2.552	2.878	3.197	3.610	3.922
19	0.688	0.861	1.066	1.328	1.729	2.093	2.205	2.539	2.861	3.174	3.579	3.883
20	0.687	0.860	1.064	1.325	1.725	2.086	2.197	2.528	2.845	3.153	3.552	3.850
21	0.686	0.859	1.063	1.323	1.721	2.080	2.189	2.518	2.831	3.135	3.527	3.819
22	0.686	0.858	1.061	1.321	1.717	2.074	2.183	2.508	2.819	3.119	3.505	3.792
23	0.685	0.858	1.060	1.319	1.714	2.069	2.177	2.500	2.807	3.104	3.485	3.768
24	0.685	0.857	1.059	1.318	1.711	2.064	2.172	2.492	2.797	3.091	3.467	3.745
25	0.684	0.856	1.058	1.316	1.708	2.060	2.167	2.485	2.787	3.078	3.450	3.725
26	0.684	0.856	1.058	1.315	1.706	2.056	2.162	2.479	2.779	3.067	3.435	3.707
27	0.684	0.855	1.057	1.314	1.703	2.052	2.158	2.473	2.771	3.057	3.421	3.690
28	0.683	0.855	1.056	1.313	1.701	2.048	2.154	2.467	2.763	3.047	3.408	3.674
29	0.683	0.854	1.055	1.311	1.699	2.045	2.150	2.462	2.756	3.038	3.396	3.659
30	0.683	0.854	1.055	1.310	1.697	2.042	2.147	2.457	2.750	3.030	3.385	3.646
40	0.681	0.851	1.050	1.303	1.684	2.021	2.123	2.423	2.704	2.971	3.307	3.551
50	0.679	0.849	1.047	1.299	1.676	2.009	2.109	2.403	2.678	2.937	3.261	3.496
60	0.679	0.848	1.045	1.296	1.671	2.000	2.099	2.390	2.660	2.915	3.232	3.460
80	0.678	0.846	1.043	1.292	1.664	1.990	2.088	2.374	2.639	2.887	3.195	3.416
100	0.677	0.845	1.042	1.290	1.660	1.984	2.081	2.364	2.626	2.871	3.174	3.390
1,000	0.675	0.842	1.037	1.282	1.646	1.962	2.056	2.330	2.581	2.813	3.098	3.300
∞	0.674	0.841	1.036	1.282	1.645	1.960	2.054	2.326	2.576	2.807	3.091	3.291

Table B-4: The Chi-Square Distribution: χ^2 Critical Values

(Table entry is the point χ^2 with given probability p lying above it.)

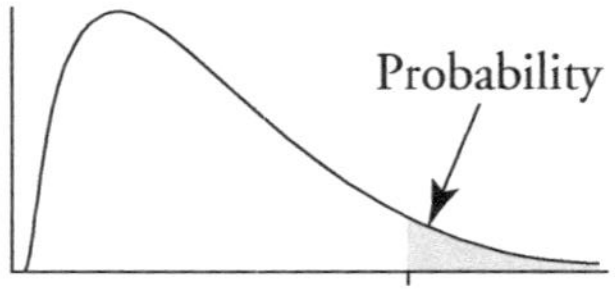

Degrees of freedom (df)	*Tail probability p* (proportion in critical region) 0.25	0.20	0.15	0.10	0.05	0.025	0.02	0.01	0.005	0.0025	0.001	0.0005
1	1.32	1.64	2.07	2.71	3.84	5.02	5.41	6.63	7.88	9.14	10.83	12.12
2	2.77	3.22	3.79	4.61	5.99	7.38	7.82	9.21	10.60	11.98	13.82	15.20
3	4.11	4.64	5.32	6.25	7.81	9.35	9.84	11.34	12.84	14.32	16.27	17.73
4	5.39	5.99	6.74	7.78	9.49	11.14	11.67	13.28	14.86	16.42	18.47	20.00
5	6.63	7.29	8.12	9.24	11.07	12.83	13.39	15.09	16.75	18.39	20.52	22.11
6	7.84	8.56	9.45	10.64	12.59	14.45	15.03	16.81	18.55	20.25	22.46	24.10
7	9.04	9.80	10.75	12.02	14.07	16.01	16.62	18.48	20.28	22.04	24.32	26.02
8	10.22	11.03	12.03	13.36	15.51	17.53	18.17	20.09	21.95	23.77	26.12	27.87
9	11.39	12.24	13.29	14.68	16.92	19.02	19.68	21.67	23.59	25.46	27.88	29.67
10	12.55	13.44	14.53	15.99	18.31	20.48	21.16	23.21	25.19	27.11	29.59	31.42
11	13.70	14.63	15.77	17.28	19.68	21.92	22.62	24.72	26.76	28.73	31.26	33.14
12	14.85	15.81	16.99	18.55	21.03	23.34	24.05	26.22	28.30	30.32	32.91	34.82
13	15.98	16.98	18.20	19.81	22.36	24.74	25.47	27.69	29.82	31.88	34.53	36.48
14	17.12	18.15	19.41	21.06	23.68	26.12	26.87	29.14	31.32	33.43	36.12	38.11

Degrees of freedom (df)	Tail probability *p* (proportion in critical region)											
	0.25	***0.20***	***0.15***	***0.10***	***0.05***	***0.025***	***0.02***	***0.01***	***0.005***	***0.0025***	***0.001***	***0.0005***
15	18.25	19.31	20.60	22.31	25.00	27.49	28.26	30.58	32.80	34.95	37.70	39.72
16	19.37	20.47	21.79	23.54	26.30	28.85	29.63	32.00	34.27	36.46	39.25	41.31
17	20.49	21.61	22.98	24.77	27.59	30.19	31.00	33.41	35.72	37.95	40.79	42.88
18	21.60	22.76	24.16	25.99	28.87	31.53	32.35	34.81	37.16	39.42	42.31	44.43
19	22.72	23.90	25.33	27.20	30.14	32.85	33.69	36.19	38.58	40.88	43.82	45.97
20	23.83	25.04	26.50	28.41	31.41	34.17	35.02	37.57	40.00	42.34	45.31	47.50
21	24.93	26.17	27.66	29.62	32.67	35.48	36.34	38.93	41.40	43.78	46.80	49.01
22	26.04	27.30	28.82	30.81	33.92	36.78	37.66	40.29	42.80	45.20	48.27	50.51
23	27.14	28.43	29.98	32.01	35.17	38.08	38.97	41.64	44.18	46.62	49.73	52.00
24	28.24	29.55	31.13	33.20	36.42	39.36	40.27	42.98	45.56	48.03	51.18	53.48
25	29.34	30.68	32.28	34.38	37.65	40.65	41.57	44.31	46.93	49.44	52.62	54.95
26	30.43	31.79	33.43	35.56	38.89	41.92	42.86	45.64	48.29	50.83	54.05	56.41
27	31.53	32.91	34.57	36.74	40.11	43.19	44.14	46.96	49.64	52.22	55.48	57.86
28	32.62	34.03	35.71	37.92	41.34	44.46	45.42	48.28	50.99	53.59	56.89	59.30
29	33.71	35.14	36.85	39.09	42.56	45.72	46.69	49.59	52.34	54.97	58.30	60.73
30	34.80	36.25	37.99	40.26	43.77	46.98	47.96	50.89	53.67	56.33	59.70	62.16
40	45.62	47.27	49.24	51.81	55.76	59.34	60.44	63.69	66.77	69.70	73.40	76.09
50	56.33	58.16	60.35	63.17	67.50	71.42	72.61	76.15	79.49	82.66	86.66	89.56
60	66.98	68.97	71.34	74.40	79.08	83.30	84.58	88.38	91.95	95.34	99.61	102.7
80	88.13	90.41	93.11	96.58	101.9	106.6	108.1	112.3	116.3	120.1	124.8	128.3
100	109.1	111.7	114.7	118.5	124.3	129.6	131.1	135.8	140.2	144.3	149.4	153.2

INDEX

G

H

I

N

O

Q

R

S

U

V

X

Z

Made in United States
Troutdale, OR
11/03/2026

56844638R00110